A TEXTBOOK OF FERTILIZERS

About The Author

Dr. Dipak Ranjan Biswas has an excellent academic carrier with undergraduate from Bidhan Chandra Krishi Viswavidyalaya (BCKV), Mohanpur, Nadia, West Bengal; Post-Graduate and Ph.D. at Indian Agricultural Research Institute (IARI), New Delhi.

Dr. Biswas is presently working as Principal Scientist, Division of Soil Science and Agricultural Chemistry, ICAR-IARI, New Delhi. He is endowed with excellent teaching skills and enjoyed teaching in Soil Science and Agricultural Chemistry. He had a long tenure of 28 years as teacher and 34 years as researchers at the ICAR-IARI, New Delhi. He guided/guiding 20 Ph.D/ M.Sc. students, most of them joined in the Agricultural Research Service (ARS) of the ICAR as Scientist. It brings together the accumulated experience and knowledge gained during his professional career on the application and management of fertilizers.

Dr. Biswas is widely known for his work in soil fertility and fertilizer use. He is the recipient of IARI Best Teacher Award for outstanding contribution and excellence in teaching; World Phosphate Institute–Fertiliser Association of India Award. His significant research contributions honoured him with many Fellowships by several National Academics and Learned Societies. He has been honoured as a Fellow of National Academy of Agricultural Sciences (NAAS), New Delhi; West Bengal Academy of Science and Technology (WAScT), Kolkata; and Indian Society of Soil Science (ISSS), New Delhi. Dr. Biswas delivered a large number of invited lectures at different scientific forum. He has published more than 200 research paper in peer reviewed journal of national and international repute including books/book chapters and conference/symposium.

Presently, Dr. Biswas is discharging his responsibilities in the national level as the Secretary, Indian Society of Soil Science (ISSS) and Editor (*Ex-Officio*) of the Journal of the Indian Society of Soil Science (JISSS), contributing significantly in shaping soil science research and education and publishing the JISSS since 2013. Dr. Biswas has been elected as the President of the Section of Agriculture & Forestry Sciences in the 108th Session (2020-2022) of the Indian Science Congress Association (ISCA), Kolkata. He is discharging his responsibilities as the Honorary Editor of Everyman's Science published by the ISCA.

Dr. Biswas's scientific contributions led to development of basic and applied nutrient management strategies to improve soil fertility and fertilizer uses; transformations of phosphorus and potassium in soils; synthesis of novel product of polymer coated controlled release formulations; solubilization of rock phosphate with organic acid loaded nanoclay polymer complex and phosphate solubilizing microorganisms as promising slow release fertilizer; bio-intervention of potassium-minerals using potassium solubilizing bacteria; development of low-cost technology to prepare phosphorus and potassium enriched compost and value added organic manures leading to crucial decision support to enhance crop production for efficient management of natural resources. Dr. Biswas also acted as the Principal Investigator in many research projects funded by the ICAR, New Delhi, as well as other organizations.

A TEXTBOOK OF FERTILIZERS

Dipak Ranjan Biswas
B.Sc. (BCKV), M.Sc. & Ph.D. (IARI)
FNASS, FISSS, FAScT
Principal Scientist
Division of Soil Science and Agricultural Chemistry
ICAR-Indian Agricultural Research Institute
PUSA, New Delhi - 110 012

NEW INDIA PUBLISHING AGENCY
New Delhi 110 034

Biswas, D.R. (2021) A Textbook of Fertilizer, New India Publishing Agency, New Delhi

NEW INDIA PUBLISHING AGENCY
101, Vikas Surya Plaza, CU Block, LSC Market
Pitam Pura, New Delhi – 110 034, India
Email: info@nipabooks.com
Web: www.nipabooks.com

For customer assistance, please contact

Phone: + 91-11-27 34 17 17 Fax: + 91-11- 27 34 16 16
E-Mail: feedbacks@nipabooks.com

ISBN : 978-93-90512-80-5

Composed and Designed by NIPA.

Dedicated to my Parents

Late Debendra Nath Biswas
and
Late Jashoda Biswas

भारतीय कृषि अनुसंधान परिषद
कक्ष क्र. 101, कृषि अनुसंधान भवन–II, पूसा, नई दिल्ली–110 012, भारत
Room No. 101, Krishi Anusandhan Bhavan-II, Pusa, New Delhi-110 012, INDIA

भाकृअनुप
ICAR

डॉ. सुरेश कुमार चौधरी / Dr. Suresh Kumar Chaudhari
उप महानिदेशक (प्राकृतिक संसाधन प्रबंधन)
Deputy Director General (Natural Resource Management)

Foreword

Fertilizer played a pivotal role in enhancing food production in India especially after the introduction of high yielding and fertilizer responsive crop varieties during the green revolution era. Fertilizer alone is considered to account for nearly one-half of this increased productivity. However, in recent years, declining soil health including widespread multi-nutrient deficiencies particularly under intensive agriculture is of prime concern of Indian agriculture. In order to make farming more profitable and sustainable, there is a need to maintain soil health and fertility through balanced and judicious fertilizer management. It may be noted that the Government is also implementing a National Mission on Soil Health Card to provide soil test based fertilizer recommendations to all the farmers in the country.

This book entitled **"A Textbook of Fertilizers"** has eight chapters covering the status of fertilizer production in India, the feedstock and raw material availability and their uses, production technology of various fertilizers including nano-fertilizers and recent technological developments. The present book is a timely contribution for updating the knowledge of teachers, scientists, researchers and students in the discipline of Soil Science and Agricultural Chemistry, Agronomy and allied disciplines.

My best compliments to the author for taking pain to write a comprehensive textbook covering all aspects of fertilizers.

(S.K. Chaudhari)

Phone: *91-11-2584 8364
Fax: *91-11-2584 8366

E-map: ddg.nrm@icar.gov.in
Website: www.icar.org.in

Preface

The book "A Textbook of Fertilizers" is the outcome of a perception by the author's 28 years of experience as a teacher to have a textbook of fertilizers for the undergraduate and postgraduate students, researchers and teachers of State and Central Agricultural Universities (SAUs/CAUs), Indian Council of Agricultural Research (ICAR) and other Academic Institutions. The courses dealing with fertilizers already occupy an important position in the curricula of all the SAUs/ CAUs, ICAR Institutes and other agricultural colleges in the country.

This publication endeavors to provide a comprehensive and authoritative treatise on fundamental and theoretical aspects of fertilizers in terms of their production technology, characterization, properties and uses under different soil-plant conditions. It emphasises the details of theories and principles and technology/ methodology for each fertilizer employed by the fertilizer industries.

This book has eight major chapters. The first chapter "Introduction", includes classification of essential nutrients and different fertilizers for crop production. Chapter 2 covers the details on nitrogenous fertilizers, their history of development of technology; different feedstock (hydrocarbon) and raw materials, commercial synthesis of ammonia, manufacturing process of different nitrogenous fertilizers and their chemical and physical properties, methods of application and fate to soils. Chapters 3 deals with the manufacturing process of different phosphatic fertilizers and their chemical and physical properties, methods of application and their fates in soils. Chapter 4 deals with the manufacturing process of different potassium fertilizers and their chemical and physical properties, methods of application and their fates in soils. Chapter 5 and 6 encompasses different fertilizers containing secondary nutrients and micronutrients, respectively. Chapter 7 deals with slow-release or controlled-release fertilizers and their importance and scope in the present context. While, specifications of different fertilizers as specified by the Fertiliser (Control) Order, 1985 (As amended up to February 2019) are given in Chapter 8.

There are several important features of this book. The purpose of bringing out this textbook of fertilizers is to provide an up to date account of fertilizers in terms of their world production and consumption *vis-à-vis* in India. This is the first attempt by the author to provide the details of production technology, the chemistry, characteristics, properties and fate of each fertilizers upon application to different soils.

In the recent years, understanding of many aspects of the commercial fertilizers as well as newly developed novel smart fertilizers including slow-release or controlled release fertilizers and nano-fertilizers have been greatly increased. In this book, the author has given systematic and in-depth coverage of the subject related to fertilizer technology and its systematic application to field crops. This book has been written to provide the fundamental aspects like basic principles and theories which are put into operation. References cited in the text are provided at the end of this book as separate chapter.

I sincerely hope that teachers and students of Soil Science in various State/ Central Agricultural Universities, ICAR institutions and other private institutions/ organizations would find this book useful in handling for best management of fertilizers and improving nutrient use efficiency under different soil-plant conditions.

I am sure that this publication will be of immense use to all concerned with fertilizers use and agricultural development such as agricultural administrators, researchers, teachers, students, soil chemists, extension workers, farmers, field advisor and marketing staff of the fertilizer industry, trading organization, training centres and international institutes.

I am very much indebted to my Professors who have taught me the subject of soil science in general, and soil fertility and fertilizer technology in particular as well as other areas of soil science. Their stimulating comments and valuable suggestions are most appreciated. I would also like to acknowledge my friends and colleagues for their continuous advice and critical comments on the subject.

Special appreciation is extended to my students Mr. Abhijit Sarkar, Mr. Kingshuk Modak and Mr. Khurshid Alam for rendering their necessary help for corrections of the manuscript as and when required. One of the most time-consuming tasks in the production of a book is that of the preparation of the manuscript. My sincere acknowledgements and appreciation for the services rendered by Mr. Vijay Kumar for his assistance in manuscript typing. I would also like to thank the Publishers for bringing out this book.

Last but not the least, I would like to acknowledge the continued encouragements and support received from my wife (Mrs. Sikha Biswas) and my son (Mr. Debanshu Biswas) during the preparation of the manuscript.

I would appreciate receiving concrete suggestions on any aspect of this publication as these would immensely help in improving the future editions of this book on fertilizers.

Dipak Ranjan Biswas

Contents

List of Acronym

ADP – Adenosine diphosphate
AG – Ammoniator-granulator
AI – Activity index
AN – Ammonium nitrate
AOAC – Association of Official Agricultural Chemists
APN – Ammonium phosphate nitrate
APS – Ammonium phosphate sulphate
APSN – Ammonium phosphate sulphate nitrate
ASP – Ammonium sulphate phosphate
ATC – 4-Amino-1,2,4-triazole hydrochloride
ATP – Adenosine triphosphate
BAAS – British Association for the Advancement of Science
BGA – Blue Green Algae
CAN – Calcium ammonium nitrate
CCC – Coated calcium carbide
CDU – Crotonylidene diurea
CEC – Cation exchange capacity
CISP – Citrate insoluble phosphate
CRFs – Controlled release fertilizers
CRH – Critical relative humidity
CSP – Citrate soluble phosphate
CWIN – Cold water (25 °C) insoluble nitrogen
DAP – Diammonium phosphate
DCP – Dicalcium phosphate
DCS – Dichlorophenyl succinamic acid

DMPP – Dimethylpyrazole phosphate

DNA – Deoxyribonucleic acid

DTPA – Diethylene triamine penta acetic acid

EDDHA – Ethylene diamine dihydroxyphenyl acetic acid

EDTA – Ethylene diamine tetra acetic acid

FACT – Fertilizers and Chemicals Travancore Limited

FCO – Fertiliser Control (Order)

FYM – Farmyard manure

GHGs – Greenhouse gases

GM – Green manure

GSFC – Gujarat State Fertilizers and Chemicals Limited

GUP – Guanylurea phosphate

GUS – Guanylurea sulphate

HA – Humic acid

HDO – Heavy diesel oil

HEDTA – Hydroxyethyle ethylene diamine triacetic acid

HFO – Heavy fuel oil

HQ – Hydroquinone

HWIN – Hot water (98-100 °C) insoluble nitrogen

IBDU – Isobutylidene diurea

IFDC – International Fertilizer Development Centre

IFFCO – Indian Farmers Fertiliser Cooperative Limited

LPG – Liquefied petroleum gas or liquid petroleum gas

MAP – Monoammonium phosphate

MCP – Monocalcium phosphate

MEA – Monoethanol amine

MOP – Muriate of potash

MPA – Metaphosphoric acid

MPSMC – Madhya Pradesh State Mining Corporation Limited

NBPT – N-(butyl) thiophosphoric triamide
NCPO – Non-catalytic partial oxidation
NFL – National Fertilizer Limited, Nangal, Punjab
NTA – Nitrilo-tri-acetic acid
NUE – Nutrient use efficiency
OPA – Orthophosphoric acid
PARP – Partially acidulated rock phosphate
PCU – Polymer coated urea
PMA – Phenyl mercury acetate
PN – Pre-neutralizer
PPA – Pyrophosphoric acid
PPCL – Pyrites, Phosphates and Chemicals Limited
PPDA – Phenyl phosphorodiamidate
PR – Phosphate rock
PSA – Phosphate solubilizing actinomycetes
PSB – Phosphate solubilizing bacteria
PSF – Phosphate solubilizing fungi
PSM – Phosphate solubilizing microorganism
PVA – Polyvinyl alcohol
RAES – Rothamsted Agricultural Experimental Station
RCF – Rhizosphere-controlled fertilizer
RCFL – Rashtriya Chemicals and Fertilizer Limited
RNA – Ribonucleic acid
RP – Rock phosphate
RSMML – Rajasthan State Mines and Minerals Limited
SAPSRPF – Superabsorbent polymer with slow-release phosphate fertilizer
SCU – Sulphur coated urea
SOM – Soil organic matter
SOP – Sulphate of potash
SRFs – Slow-release fertilizers

SSP – Single superphosphate
ST – Sulphathiazole
TAP – Triammonium phosphate
TCA – Tricarboxylic acid
TEA – Triethanol amine
Tetra-PPA – Tetra-polyphosphoric acid
Tri-PPA – Tri-polyphosphoric acid
TSP – Triple superphosphate
TVA – Tennessee Valley Authority
UAP – Urea ammonium phosphate
USEPA – United States Environment Protection Authority
USG – Urea super granule
WBMDTC – West Bengal Mineral Development and Trading Corporation Limited
WISP – Water insoluble P
WSP – Water soluble P

Chapter 1

Introduction

Plant absorbs large number of elements from the soil and other sources during their growth and development. But all these elements are not essential for plant growth. Seventeen mineral elements are known to be essential for the growth and development of all plants. A list of essential nutrients, their source and forms by which they are absorbed by plants are presented in Table 1.1.

Except carbon (C), hydrogen (H) and oxygen (O), all the other nutrients are obtained by plants largely from soils, fertilizers or manures. It has been seen that the number of nutrients obtained from soil is about 5 times the number of nutrients obtained from air and water. However, most of the plant tissue, in general, from 74 to 99.5% is made up of C, H and O *i.e.* nutrients obtained from air and water and only 0.5 to 26% of the plant tissue is synthesized from soil constituents. In ordinary (normal) conditions of agriculture, except drought, C, O and H are not limited for plant growth. It is the remaining 14 elements from soil that are usually limiting plant development and hence considerable emphasis is placed on the improvement of soil in respect of greater availability of these 14 nutrients. It may be borne in mind that all nutrients are essential for all crops even though their requirements differ from crop to crop and even among varieties of a given crop. Secondly, all nutrients are equally essential regardless of amounts required. For example, to produce one tonne of grain, wheat may absorb 25, 9, 33, 3.7, 5.3 and 4.7 kg of nitrogen (N), phosphorus (P), potassium (K), sulphur (S), calcium (Ca) and magnesium (Mg), respectively and 56, 624, 70, 24, 48 and 2 g of zinc (Zn), iron (Fe), manganese (Mn), copper (Cu), boron (B) and molybdenum (Mo), respectively. If the crop (any crop) does not get the smallest amounts of micronutrients needed, it cannot make the best use of abundant NPK.

According to **Liebig's Law of minimum** which he enunciated in 1855, the growth of a plant is limited by that nutrient which is in shortest supply. This law in fact lays the ground rules for balanced crop nutrition.

Table 1.1. Essential nutrients, their discoveries, average concentration in plant tissues and forms by which they are absorbed by plants

Nutrients	Essentiality discovered	Average conc. in plant tissue	Forms absorbed by plants
Hydrogen (H)	Since time immemorial	6%	H_2O
Oxygen (O)	Since time immemorial	45%	O_2, H_2O
Carbon (C)	Priestley *et al.* (1800)	45%	CO_2
Nitrogen (N)	Theodore de Saussure (1804)	1.5%	NH_4^+, NO_3^-
Phosphorus (P)	C. Sprengel (1839)	0.2%	$H_2PO_4^-$, HPO_4^{2-}
Potassium (K)	C. Sprengel (1839)	1.0%	K^+
Sulphur (S)	Sachs and Knop (1860)	0.1%	SO_4^{2-}
Calcium (Ca)	C. Sprengel (1839)	0.5%	Ca^{2+}
Magnesium (Mg)	C. Sprengel (1839)	0.2%	Mg^{2+}
Iron (Fe)	E. Gris (1843)	100 mg kg^{-1}	Fe^{2+}
Manganese (Mn)	J.S. McHargue (1922)	20 mg kg^{-1}	Mn^{2+}
Zinc (Zn)	A.L. Sommer and C.P. Lipman (1926)	20 mg kg^{-1}	Zn^{2+}
Copper (Cu)	A.L. Sommer, C.P. Lipman and G. McKinney (1931)	6 mg kg^{-1}	Cu^+, Cu^{2+}
Boron (B)	K. Warington (1923)	20 mg kg^{-1}	H_3BO_3, $H_2BO_3^-$, HBO_3^{2-}, BO_3^{3-}
Molybdenum (Mo)	D.I. Arnon and P.R. Stout (1939)	0.1 mg kg^{-1}	MoO_4^{2-}
Chlorine (Cl)	T.C. Broyer, A.B. Carlton, C.M. Johnson and P.R. Stout (1954)	100 mg kg^{-1}	Cl^-
Nickel (Ni)	P.H. Brown, R.M. Stout and E.E. Cary (1987)	0.1 mg kg^{-1}	Ni^+, Ni^{2+}

According to **Arnon and Stout (1939)** an element is essential for plant growth if,

(a) The plant is unable to grow normally or complete its life cycle without it;

(b) The need of such a nutrient is specific and its deficiency cannot be prevented or corrected by adding some other nutrients; and

(c) The nutrient plays a direct role in the plants active (metabolic) processes and meets its nutritional needs.

A somewhat more flexible definition of essentiality has been proposed by J.D. Nicholas, who coined the term "***functional*** or ***metabolic nutrient***" which would include any mineral element involved whether or not it produced specific action. If this definition is followed then silicon (Si), sodium (Na) and vanadium (V) could also be classified as essential plant nutrients.

Macronutrients: Out of 17 elements, six elements, namely, N, P, K, S, Ca and Mg are used by plants in large quantities. Since these nutrients are utilized in relatively large amounts, they are designated as **macronutrients**. The first three nutrients namely, N, P and K are utilized by plants in considerable quantities and hence often called as **major nutrients**. The fertilizer industry has been supplying these three nutrients in large quantities, in the form of commercial fertilizers to the farmers. The remaining three macronutrients (S, Ca and Mg) are called **secondary nutrients** due to their secondary importance to the manufacturers of N, P and K fertilizers. For example, single superphosphate (SSP) contains both Ca and S. Similarly, ammonium sulphate also supply S besides N.

Micronutrients: The remaining eight nutrients (Fe, Zn, Cu, Mn, Mo, B, Cl and Ni) are used by crops in very small quantities. These are called **micronutrients.** Other synonyms given to micronutrients are **trace, minor** or **rare elements**. The micronutrients, however, are as essential as macronutrients.

With the emergence of deficiencies of the nutrients other than N, P and K, carriers of S, Zn, Mn, Fe, Mo, B as fertilizers are gaining importance. Recently, there has been a growing need for the inclusion of secondary and micronutrients in fertilizers or nutrients other than the primary elements (N, P and K). There are several reasons for this.

a) ***High analysis NPK Fertilizers:*** Firstly, there is an increasing trend of use of high analysis NPK fertilizers in order to decrease the handling, transportation and application costs. As a result there is no room for the secondary and micronutrients that formerly were included incidentally because they were present in the original raw materials.

b) ***Cropping intensity:*** Secondly, with increasing cropping intensity, the soil nutrients are depleting. For sustainable crop production and maintenance of soil health, the nutrients which are mined through crop removal must be replenished through external sources. Thus, in addition to macronutrients, application of secondary and micronutrients are becoming more important.

c) ***Crop requirement:*** Some crops require more secondary and micronutrients for their growth and development. For example, pulses and oilseeds require more amount of S as it is an important component of amino acids.

d) ***Problem soils (acid or alkaline soils):*** In addition to serving important functions as plant nutrients, secondary nutrients are also used extensively as soil amendments to correct undesirable soil conditions. Maintenance of the proper soil pH, generally 6-7, is most important in order to obtain efficient utilization of nutrients by the plants. Certain soils are naturally productive; however, to retain their high productively, the addition of the amendments is necessary. In soils that are naturally unproductive because of excessive acidity or alkalinity, these conditions must be corrected before the soils can become economically productive.

1.1. Fertilizers

Fertilizers are the material containing one or more of the essential plant nutrients required for plants growth and development. It may be defined as any solid, liquid or gaseous substances of definite chemical composition and high analytical value. They are also known as **chemical fertilizers**. Fertilizers occur either as a natural deposit or are synthesized by manufacturer in a chemical factory. When fertilizers are applied to soil, they undergo chemical transformations before releasing the nutrients in plant available form. Upon application, fertilizer materials dissolve fully or sparingly in the soil-water system and release the nutrient(s) in the ionic form which are then absorbed by the plants. Irrespective of the mode of release, a nutrient has to be present in a specific ionic form, so that plants can absorb it from soil solution through the root system.

1.2. Manures

Manures are also the materials containing essential plant nutrients. However, they are plant and animal origin having lesser concentration of plant nutrients compared to fertilizers. They are known as **organic fertilizers**.

1.3. Classification of Fertilizers

Fertilizers may be classified according to their presence of number of major nutrient *i.e.* whether they provide a single nutrient (only N, P or K) or two or more major nutrients. Based on this, fertilizers can be grouped into three categories: (i) Straight fertilizers, (ii) Compound or complex fertilizers, and (iii) Mixed fertilizers or Mixtures. A schematic diagram showing various fertilizers is presented in **Figure 1.1.**

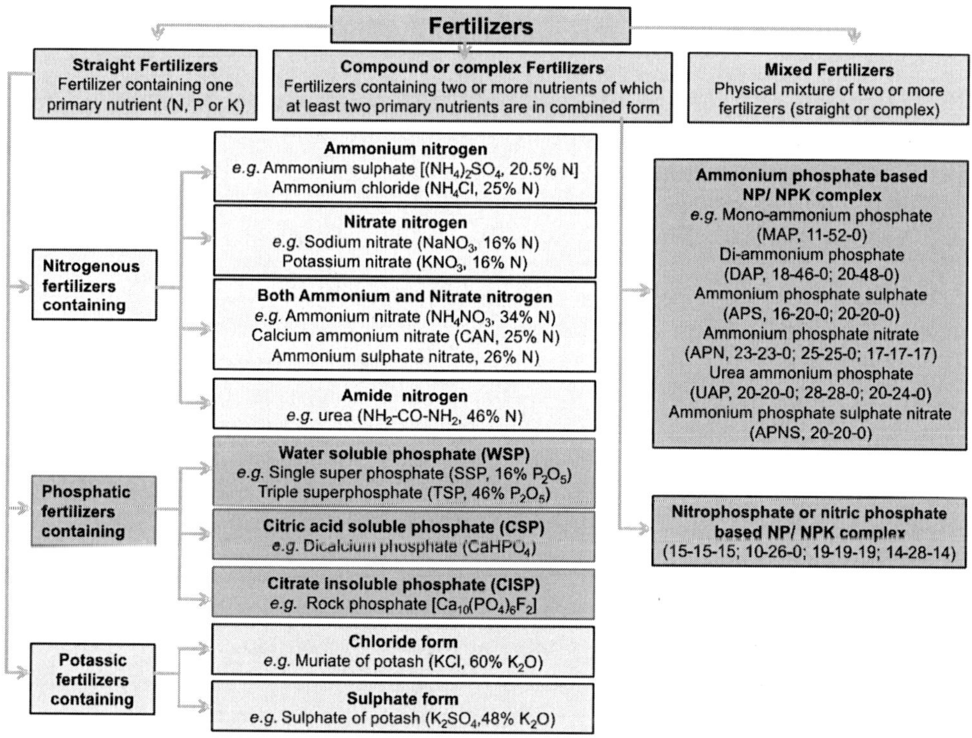

Figure 1.1. Schematic diagram of different fertilizers

(i) Straight fertilizers: Straight fertilizers are those fertilizers which contain only one primary nutrient element (N, P or K). Examples of straight fertilizers are ammonium sulphate (supplying only N as major nutrient), urea (supplying only N as major nutrient), single superphosphate (supplying only P as major nutrient), *etc*. Straight fertilizers may be (a) **straight nitrogenous fertilizers** which supply N as the major nutrient like urea, ammonium sulphate, ammonium chloride, ammonium nitrate, calcium ammonium nitrate, *etc.*; (b) **straight phosphatic fertilizers** which supply P as the major nutrient like single superphosphate, triple superphosphate, *etc.*; and (c) **straight potassic fertilizers** which supply K as the major nutrient like muriate of potash, sulphate of potash, *etc.*

(ii) Compound or Complex fertilizers: Compound or complex fertilizers are those containing two or more nutrients of which at least two primary nutrients are in chemically combined form. Complex fertilizers are also called as **Multinutrient fertilizers**. Examples of complex fertilizers are monoammonium phosphate or diammonium phosphate (supplying only

N and P as major nutrients), NPK complex fertilizer (supplying only N, P and K as major nutrients), *etc.*

(iii) Mixed fertilizers or mixture: Mixed fertilizers or mixture are those which are prepared by physically mixing two or more straight fertilizers or two or more complex fertilizers. Fertilizer mixtures are usually prepared to meet specific needs of a crop in a region. These are prepared by mixing two fertilizers just before application in the field.

Fertilizers are also sometimes classified as **inorganic** and **organic**. Inorganic fertilizers exclude carbon-containing materials except urea. Organic fertilizers are usually the plant or animal-derived matter. Inorganic fertilizers are sometimes called **synthetic fertilizers** because of various chemical treatments involved in their manufacture.

In addition to above categories, fertilizers may also be called as **complete fertilizer** which contains all the three major nutrients such as nitrogen (N), phosphate (P_2O_5) and potash (K_2O).

Depending on the form, nitrogenous fertilizers may be either **ammonium fertilizers** (*e.g.* ammonium sulphate, ammonium chloride, *etc.*) or **nitrate fertilizers** (*e.g.* potassium nitrate, sodium nitrate, *etc.*) or **both ammonium and nitrate fertilizers** (*e.g.* ammonium nitrate, calcium ammonium nitrate, *etc.*) or **amide fertilizer** (urea).

The forms of P present in superphosphate, ammonium polyphosphate and ammonium metaphosphate are orthophosphate, polyphosphate and metaphosphate, respectively. Thus, they are termed **orthophosphate fertilizers, polyphosphate fertilizers** and **metaphosphate fertilizers**, respectively. Again, the phosphatic fertilizers may be termed as **water soluble phosphates** (*e.g.* superphosphates, ammonium phosphates, *etc.*) or **citrate soluble phosphates** which is not soluble in water but, soluble in neutral 1 *N* ammonium citrate solution (*e.g.* dicalcium phosphate, nitrophosphates, *etc.*) or it may be **citrate insoluble phosphates** (*e.g.* rock phosphate which is not soluble in water as well as neutral 1 *N* ammonium citrate solution).

Fertilizers can also be categorised as **acidic fertilizers** as upon application to soil they undergoes transformation and induce acidity to soil. Examples of acidic fertilizers are ammonium and amide containing N fertilizers. While, sodium nitrate, upon application to soil, induces alkalinity. Thus, it is categorised as **alkaline fertilizer**. On the other hand, calcium ammonium nitrate is categorised as **neutral fertilizer** as it do not induce acidity or alkalinity upon application to soils.

Following terms are commonly used in mixed fertilizers:

(a) **Fertilizer grade:** Fertilizer grade refers to the guaranteed minimum percentage of nitrogen (N), phosphate (P_2O_5) and potash (K_2O) present in the fertilizer mixture. Fertilizer grade is also called as NPK ratings which consist of three numbers separated by dashes (*e.g.*, 15-15-15 or 20-24-10) describing the chemical content of fertilizers. The first number represents the percentage of N in the product; the second number, P_2O_5; and the third, K_2O. Fertilizers do not actually contain P_2O_5 or K_2O, but the system is a conventional shorthand for the amount of the phosphorus (P) or potassium (K) in a fertilizer. Thus, a fertilizer of 100 kg with a grade 20-24-10 contains a minimum of 20 kg of N, 24 kg P_2O_5 and 10 kg of K_2O, respectively.

(b) **Fertilizer ratio:** This refers to the ratios of the percentages of nitrogen (N), phosphate (P_2O_5) and potash (K_2O) in the fertilizer mixture, *e.g.* the fertilizer grade 12-6-6 has a fertilizer ratio of 2:1:1.

(c) **Conditioners:** These are materials which are added to fertilizer mixtures during their preparation for reducing the hygroscopic activity and improving their physical condition of the fertilizer product. Such materials are low-grade organic materials.

(d) **Filler:** These are the weight makeup materials added to fertilizer ingredient to produce a mixture of the desired grade. The filler materials may be sand, soil, china clay, coal powder, *etc.*

Chapter 2

Nitrogenous Fertilizers

Nitrogen (N) is an essential element for all forms of life because it is a major component of amino acids, which are the building blocks of proteins, nucleic acids such as deoxyribonucleic acid (DNA) and other cellular constituents. It is a structural component of chlorophyll in plants, which is used in photosynthesis to make their food. Nitrogen ranks first among the essential nutrient elements required for plants. Its careful management is very important for crop production.

Nitrogen exists as two main stable isotopes (isotopes of a given element have the same number of protons but different numbers of neutrons in each atom) namely, $^{14}{}_{7}N$ (abundance 99.63%) and $^{15}{}_{7}N$ (abundance 0.37%). These can be separated by chemical exchange or by thermal diffusion. Both $^{14}{}_{7}N$ and $^{15}{}_{7}N$ isotopes of N are *stable isotope* which do not disintegrate as in case of other radioactive isotope like $^{32}_{15}P$. There are four other isotopes of N *viz.* $^{12}{}_{7}N$, $^{13}{}_{7}N$, $^{16}{}_{7}N$ and $^{17}{}_{7}N$, mostly are radioactive having very short half-life and not used in agriculture. The $^{13}{}_{7}N$ has a half-life of about 10 min, while other three have half-life of only few sec. The $^{15}{}_{7}N$ (tracer) isotope is used as a tool for determination of pathways of transformation of N in soil or plant or animal.

The major part of N comes from atmosphere (air). On the basis of volume by volume (v/v), air contains about 78-79% N_2, 20-21% oxygen (O_2) and 0.03% carbon dioxide (CO_2). Most living organisms cannot utilize N_2 gas present in the atmosphere directly because it is present as *di-nitrogen* form or *di-atomic* form (N≡N). It is called as an inert gas. Therefore, to break this triple bond, huge amount of energy is required. After breaking of triple bond of N_2 molecules, it combines with other element particularly with O_2 or hydrogen (H) to form oxides of N (NO, NO_2, NO_3 and N_2O) or ammonia (NH_3). Thus, breaking of triple bond is must for forming NH_3 which further can be used as the key material for almost all fertilizers. Therefore, the fixation of N_2 is vital for production of all nitrogenous and complex fertilizers like NP/NPK and nitrophosphates.

In nature, two principal processes of N_2-fixation are known which helps in breaking the triple bond of N_2 ($N{\equiv}N$). These are (i) Biological N_2-fixation or symbiotic N_2-fixation, and (ii) Chemical N_2-fixation. Biological N_2-fixation or symbiotic N_2-fixation has certain limitations as it is restricted to only legume crops. *Rhizobium*, *Azotobacter*, Blue Green Algae (BGA), *Azolla* have the ability to fix high amount atmospheric N_2 into plant available form. The process of inducing the molecular N_2 or di-nitrogen into compounds of nitrogen is known as chemical fixation of N. In the chemical process, high amount of energy is required while biological process requires very less amount of energy.

2.1. History of Development of Technology for Nitrogen Fertilizer

Although an adequate supply of N_2 is available in the atmosphere all over the world, it must be converted to a fixed form before it is suitable for plant use. There are some industrial processes which have been developed for fixation of atmospheric N_2. These are discussed below:

2.1.1. Electric-Arc Process for the Production of Nitric acid

It is also called as ***Birkeland-Eyde*** process. The **Birkeland–Eyde process** was one of the competing industrial processes in the beginning of N-based fertilizer production. It was developed by **Norwegian** industrialist and scientist **Prof. Kristian Birkeland** along with his business partner **Engineer Samuel Eyde** in 1903 (Birkeland, 1906; Eyde, 1909). In fact, the Birkeland-Eyde process is the first plasma NO synthesis process of N_2 fixation by chemical means ever realized successfully on the industrial scale. This process was used to fix atmospheric N_2 into nitric acid (HNO_3), one of several chemical processes generally referred to as N-fixation. The final product of this process was HNO_3, produced by first fixing atmospheric N_2 and O_2 into nitric oxide (NO), then absorbing it in water. The resultant HNO_3 was then used as a source of nitrate (NO_3^-) in the reaction. Based on this process, they started a company, named the **Norwegian Hydro Electric Nitrogen Company (NorskHydro)**, producing nitrogen compounds using electricity particularly electric arc technology.

Birkeland and Eyde observed that during lightning, atmospheric N_2 and O_2 combine together using the energy formed during lightning and forms NO (oxides of nitrogen). This NO is then reacted with H_2O to form nitrite (NO_2) and finally HNO_3. The HNO_3 thus formed during lightning dissolves and comes to Earth with rain as a very dilute solution. In course of time, it becomes part of the N of the soil and thus N content of cultivated soil is enriched.

$$\underset{\text{(Air)}}{N_2 + O_2} \xrightarrow[\text{(3000-3500 °C)}]{\text{Lightning or Electric arc}} 2\,NO + \Delta H \quad \ldots(2.1)$$

$$2\,NO + O_2 \rightarrow 2\,NO_2 \quad \ldots(2.2)$$

$$3\,NO_2 + H_2O \rightarrow 2\,HNO_3 + NO \quad \ldots(2.3)$$

Prof. Birkeland and Eyde developed the technique to fix atmospheric N_2 by passing air through an electric arc at 3000-3500 °C to produce nitrogen oxides (NO). They converted N_2 gas into NO by reacting it with O_2 in presence of electric are which also produce very high energy as in case of lightning, but it was soon abandoned because of large power requirement and high capital cost involved. The process is economic only if cheap hydroelectricity is available. The Birkeland–Eyde process is also relatively inefficient in terms of energy consumption. Therefore, in the 1910s and 1920s, it was gradually replaced by a combination of the Haber process and the Ostwald process.

In the Birkeland-Eyde process, an electrical arc is formed between two copper tube electrodes powered by a high voltage alternating current of 5kV. The plasma temperature in the disc is in excess of 3000 °C. Air is blown through this arc, causing some of the N_2 to react with O_2 forming NO. By carefully controlling the energy of the arc and the velocity of the air stream, yield of up to approximately 4-5% NO can be obtained at 3000 °C and less at lower temperatures. The process is extremely energy intensive. Birkeland used a nearby hydroelectric power station for the electricity. The same reaction is carried out by lightning, providing a natural source for converting atmospheric N_2 to soluble nitrates. The hot NO is cooled and combines with atmospheric O_2 to produce NO_2. This NO_2 is then dissolved in water to give rise to HNO_3, which is then purified and concentrated by fractional distillation.

2.1.2. Cyanamide Process

The cyanamide process is also known as **Frank and Caro** method. This method was developed in Germany by chemists **Prof. Adolph Frank** and **Dr. Nikodem Caro** between 1895 and 1898. The Frank and Caro process was the first commercial process that was used worldwide to fix atmospheric N_2 and the product was used as fertilizer known as **lime-nitrogen** or **nitrolim**. Nitrogen was fixed in the form of calcium cyanamide by reacting calcium carbide with pure N_2. The cyanamide process grew rapidly, reaching its peak in 1918. However, this process was very energy intensive.

In this process, atmospheric N_2 is fixed by the reaction of calcium carbide (CaC_2) with N_2 gas in a reactor vessel at about 1000 °C. The reaction is exothermic and self-sustaining once the reaction temperature is reached. Originally, the reaction took place in large steel cylinders with an electrical resistance element providing initial heat to start the reaction but modern production uses rotating ovens. The synthesis produces a solid mixture of calcium cyanamide ($CaCN_2$) and carbon or coke (C).

When limestone ($CaCO_3$) is heated along with coke or coal (C) at a very high temperature in presence of atmospheric N_2 calcium cyanamide is produced. The fundamental reactions of involved are given below:

$$CaCO_3 \xrightarrow[\text{(1100-1300 °C)}]{\text{ignited}} CaO + CO_2 \quad \text{...(2.4)}$$

$$CaO + 3\ C\ (\text{Coke}) \xrightarrow[\text{(2200 °C)}]{} \underset{\text{(Calcium carbide)}}{CaC_2 + CO} \quad \text{...(2.5)}$$

$$CaC_2 + N_2 \xrightarrow[\text{(1000-1100 °C)}]{} \underset{\text{(Calcium cyanamide)}}{CaCN_2 + C} \quad \text{...(2.6)}$$

Compound calcium cyanamide ($CaCN_2$) is perfectly white in colour. When $CaCN_2$ is treated with steam at 110 to 115 °C and 6 atoms pressure, NH_3 is formed.

$$CaCN_2 + 3\ H_2O \xrightarrow[\text{(110-115 °C)}]{} 2\ NH_3 + CaCO_3 \quad \text{... (2.7)}$$

The compound $CaCN_2$ is used as fertilizer. But, because of hydrolysis, it gives $Ca(OH)_2$ and NH_3.

$$CaCN_2 + 4\ H_2O \xrightarrow[\text{(Hydrolysis)}]{} Ca(OH)_2 + 2\ NH_3 + CO_2 \quad \text{... (2.8)}$$

The raw materials required for Frank-Caro method are coal and limestone which are cheap and abundant. This process has a distinct advantage over the arc process in its one-fourth electric energy requirement. Still, it involves a huge amount of electricity and $Ca(OH)_2$ is highly alkaline.

In its first decades, the world market for inorganic fertilizer was dominated by factories utilizing the cyanamide process. The cyanamide industry developed rapidly during World War-I because cyanamide N can be converted to NH_3 then to HNO_3 which can acts as explosive materials.

The first full-scale factories were established in 1905 in Italy and Germany. From 1908, the Frank–Caro process was used at North Western Cyanamide Company at Odda, Norway which was by far the largest in the world. The cyanamide factory at Odda ceased operation in 2002. It is still intact and is a Norwegian candidate to the UNESCO World Heritage List.

2.1.3. Serpek Process

Another way to convert atmospheric N_2 gas was by **Serpek process** given by **Ottokar Serpek** (1890) in Norway worked with nitrides of aluminium. Here, N_2 gas is reacted with certain metals of group II (*e.g.* Mg) and III (Bo, Al) in the periodic table. When N_2 gas is passed through these heated metal it reacts with the metal, and form NH_3 as shown below:

$$\underset{\text{(hot)}}{2\,Al + N_2} \longrightarrow \underset{\text{(Aluminium nitride)}}{2\,AlN} \xrightarrow[+3\,H_2O\,\text{(steam)}]{\text{hydrolysed}} Al(OH)_3 + NH_3 \quad \ldots (2.9)$$

$$\text{or,}\quad 2\,Al + N_2 \longrightarrow 2\,AlN\ \text{(Aluminium nitride)} \quad \ldots(2.10)$$

$$AlN + \underset{\text{(steam)}}{3\,H_2O} \xrightarrow{\text{Hydrolysis}} Al(OH)_3 + NH_3 \quad \ldots(2.11)$$

Serpek process is an indirect process of synthesizing NH_3 because hydrogen gas (H_2) comes from steam indirectly. At present, the Serpek process is also not used to produce NH_3. The electric arc process, cyanamide process and Serpek process become obsolete because they could not compete economically with the process for synthesizing NH_3 by other methods.

2.1.4. Haber-Bosch Process for Ammonia Synthesis (Fritz Haber and Carl Bosch, 1913)

The production of NH_3 by reacting H_2 with atmospheric N_2 is the basis for N-fertilizer industry today. The process of combining atmospheric N_2 with H_2 known as **Haber-Bosch** process was developed in 1908 and commercialized in 1913, eventually took over the above mentioned processes. The main reasons were that the comparatively less energy consumption and could produce higher volumes of NH_3. The first full-scale NH_3 plant was set up in 1913 by **Fritz Haber** and **Carl Bosch** at **Oppau**, Germany. The process is known as Haber-Bosch process.

$$N_2 + 3\,H_2 \rightarrow 2\,NH_3 + \Delta E\ \text{(heat)} \quad \ldots(2.12)$$

During the early years of Haber-Bosch process, the yield of this process was only 5-6%. The yield of NH_3 can be increased by following the important laws or principles involved in the above reaction namely, (i) Law of Mass Action; and Law of Chemical Equilibrium (Le Chatelier's Principle, 1884).

(i) Law of Mass Action (Cato M. Guldberg and Peter Waage, between 1864 and 1879): *This law states that the rate of any chemical reaction is directly proportional to the product of the activities or concentrations of the reactants.* Guldberg and Waage also recognized that chemical equilibrium is a dynamic process in which rates of reaction for the forward and backward reactions must be equal at chemical equilibrium.

(ii) Le Chatelier's Principle (1884): Every system tries to attain in equilibrium. In a chemical reaction, chemical equilibrium is the state in which the forward reaction rate and the reverse reaction rate are equal. The result of this equilibrium is that the concentrations of the reactants and the products do not change. *Le Chatelier's Principle states that when any system or dynamic equilibrium is disturbed by changing its conditions like temperature, volume or pressure, the system will shift itself to attain its new equilibrium or it will nullify the extra change and in such conditions, the position of equilibrium moves to partially reverse the change.* In chemistry, the Le Chatelier's Principle is used to manipulate the outcomes of reversible reactions, often to increase the yield of reactions.

According to Le Chatelier's Principle, the yield of NH_3 production may be increased by following the conditions/requirements:

(i) The reaction should be performed in low temperature since the reaction is exothermic. However, a temperature range from 400-450 °C is optimum.

(ii) The reaction should be done at high temperature (200-1000 kg cm^{-2} atmospheric pressure).

(iii) Molecular ratio of N_2:H_2 should be 1:3.

(iv) Purity of gases should be as pure as possible. About 80% of the cost goes for the purification of the gases.

In Haber-Bosch method of synthesis of NH_3, the two gases namely, N_2 and H_2 reacts directly and thus it is called direct method. All the other methods are called indirect method.

Although several other processes have been developed since 1913, the more energy efficient Haber-Bosch process gradually took over other processes because of the following reasons:

(a) the method is easy for preparation of synthesis of NH_3 gas;

(b) the purification of synthesis gas is simple;

(c) the design of the NH_3 converter is better; and

(d) the method for recovering NH_3 from the converter effluent gas is higher.

2.2. Raw materials for synthesis of ammonia

For manufacturing of ammonia (NH_3), all processes utilize atmospheric air as the source for nitrogen gas (N_2) which is almost free of cost. Whereas, hydrogen gas (H_2) can be produced from almost any hydrocarbon or carbonaceous materials called feedstock. However, careful consideration must be given to the choice of raw materials for economic cost for producing H_2. Thus, production of H_2 gas is the challenging aspect in synthesis of NH_3. Ideally, the feedstock should be cheap. Prior to 1945, coal or coke was the major raw material used for the production of H_2 required for NH_3 synthesis. After 1945, there has been continuing trend towards the use of petroleum products like natural gas, or naphtha or heavy fuel oil (HFO) as the feedstock materials throughout the world. The manufacturing plant also designed accordingly. Since there was very little amount of natural gas in India during seventies and eighties, naptha was the major H_2 feedstock. However, at present natural gas is the major feedstock in India because huge amount of natural gas is available.

2.2.1. Natural gas

Natural gas is present in various regions over the world. In India, it comes from the oil wells such as in Bombay high, Gujarat, Assam and other refineries. Its composition varies depending on the geographical location and the type of deposits. Natural gas contains mainly methane (CH_4) mixed with higher hydrocarbons like ethane (C_2H_6), propane (C_3H_8), butane (C_4H_{10}), *etc.* It contains 60-96% CH_4, 4-60% C_2H_6 and higher hydrocarbon (C_nH_{2n+2}). The crude gas may contains dust and gaseous impurities like hydrogen sulphide (H_2S), carbon dioxide (CO_2), carbon monoxide (CO) and other S-bearing compounds. These impurities affect H_2 production and must be removed before the gas is suitable for use as the feedstock for NH_3 synthesis. Sulphur is also poisonous to the feedstock and it will reduce the production of H_2.

The major advantage of natural gas as the feedstock is the high ratio of H:C. Higher is the H:C ratio in the hydrocarbon, more is the advantage. In case of CH_4, the ratio of H:C is 4:1; whereas the ratios of H:C for C_2H_6, C_3H_8 and C_4H_{10} are 3:1, 8:3 (2.67:1) and 10:4 (2.5:1), respectively. Since, H:C ratio is very high in CH_4, therefore, it is easier to convert the hydrocarbon into gaseous H_2. In the NH_3 synthesis process, all the carbon (C) present in natural gas is subsequently converted to CO or CO_2, and must be removed in the gas purification step prior to NH_3 synthesis. Thus, if C is les compared to H (as in case of CH_4) it will be easy to remove this C as CO or CO_2.

Natural gas can be classified as (i) Associated and (ii) Non-associated. Associated natural gas occurs with crude oil and can be separated from the oil when pressure is reduced. The gas is present within the solid-solution equilibrium. In case of non-associated natural gas, it is present on the surface of the oil wells which is in the free states or uncombined form. Thus, the gas comes directly from the oil wells.

2.2.2. Naptha

Napth is obtained from the liquid fraction of petroleum. It is the second important feedstock in India now-a-days. Gradually, use of naptha as feedstock is decreasing because of the high cost of petroleum product. Naptha is produced during the refining of crude oil. It is preferred as a feedstock in areas where natural gas is not readily available.

The properties of naptha are:

(i) It is a clear liquid.

(ii) The boiling range of naptha is 41-121 °C.

(iii) It is lighter than water. It's specific gravity is 0.669 g cm^{-3}.

(iv) Chemical composition of naptha shows that basically it consists of hydrocarbon *i.e.* pentane plus (C_5H_{12}) hydrocarbon (>96.7%) which contain C_5 and above (C_5H_{12}, C_5H_{12}, *etc.*). Thus, it has low H:C ratio as compared to CH_4. Therefore, naptha is less advantageous than CH_4 (natural gas). Elemental-wise it contains about 80% C, 12% H, 3.5% S, and 3% H_2S and others gases. The S content in naptha is very high compared to natural gas.

2.2.3. Heavy Fuel Oil (HFO)

Heavy fuel oil (HFO) is also known as heavy diesel oil (HDO). The HFO is produced by partial oxidation of fuel oil (say, diesel). It is a high energy content material. Elemental-wise, it contains 84% C, 11% H, 3.5% S, and 1.5% H_2S and others. Presence of S somehow will lower the production of H_2. Thus, S should be removed before gasification. Otherwise, these S-compounds will form some poisonous material. Nickel (Ni) may present and if Ni is combined with S, the poisonous compound will form. The major disadvantages of using HFO are (i) its S contents (3.5%) and (ii) handling problem. The HFO is having high viscosity. Heat must be given to keep the HFO in a liquid form (state) and to reduce viscosity. During winter, it becomes solid and thus transportation is a problem.

2.2.4. Coal or Coke

It is a major source of hydrocarbon. This is used in metalargy as energy source. Since coal is available at Raniganj (West Bengal) and its surrounding areas, it is used widely in plants near these areas such as Ramagundam (Telangana), Talcher (Odisha) and Korba (Madhya Pradesh). However, coal is less important as feedstock for NH_3 synthesis because it cannot compete economically with liquid or gaseous petroleum products. Moreover, good quality coke is used in metallurgical purposes. The processes required for coal gasification and plant investment, maintenance and operating costs are higher.

Coal and coke as such does not contain any H_2. But it contains pyridine bases (C_5H_5N) which when converted by destructive distillation (distillation in a closed system *i.e.* distillation in absence of air so that oxidation does not occur) to form several H-containing compounds. This is called ***"Coal gas"***. It is about 17% of the total weight of the coal. After distillation, the upper liquid portion is called ***"ammoniacal liquor"*** and bottom layer comprises of semi-solid material, known as ***"tar"***. This can serve as the energy source by breaking up its H-bond.

2.2.5. Liquid Petrolium Gas

Liquefied petroleum gas or liquid petroleum gas (LPG) is a flammable mixture of hydrocarbon gases containing mainly propane (C_3H_8) and butane (C_4H_{10}). It also contains pentane (C_5H_{12}) and higher hydrocarbon (C_nH_{2n+2}). It is separated by liquefaction from natural gas or from refinery gas. Therefore, the principal sources of LPG are oil refineries and natural gas processing plants. The LPG is used as fuel in heating appliances, cooking equipments and vehicles. It is also used as an aerosol propellant and a refrigerant, replacing chlorofluorocarbons in an effort to reduce damage to the ozone layer. When specifically used as a vehicle fuel it is often referred to as auto gas. As far as production of NH_3 is concerned, it has not yet became an important raw material.

2.2.6. Electrolysis of H_2O

This is a very simple process for producing H_2. When electricity is passed through water, pure H_2 can be produced.

$$H_2O \xrightleftharpoons{\text{Electrolysis}} \underset{\substack{\downarrow \\ H_2}}{H^+} + \underset{\substack{\downarrow \\ O_2}}{OH^-} \quad \text{...(2.13)}$$

The mixture of gases produced contains H_2 and O_2. This method produces H_2 gas which is almost pure. The major problem of this method is very high cost of electricity. But in NFL (National Fertilizer Limited, Nangal, Punjab) plant, the H_2 is produced by this method. This is because it is situated near the Bhakra Nangal Dam in Himachal Pradesh from where hydroelectricity is obtained with a cheaper cost. The advantage of this method is that heavy water (D_2O) may be obtained by electrolysis, provided it is subjected to a series of electrolysis ($H_2O \rightarrow D_2O$). However, the cost of production of D_2O is very high.

2.2.7. By-product H_2

By-product H_2 is obtained from several industries. These are:

(i) NaCl plant: During the extraction of Na from NaCl, some H_2 is produced as by-product.

(ii) Coke-oven gas: The most important source of by-product H_2 is coke-oven gas. The gas is rich in H_2 (~50%) and contains some NH_3 which can be recovered to produce $(NH_4)_2SO_4$ or aqua NH_3 or ammonium phosphate.

(iii) Chlorine plant

(iv) Caustic plant

(v) Petroleum refineries

(vi) Plants producing ethylene, acetylene and butadene.

(vii) Bio-gas plant also contains about 50-60% CH_4 and can be used as source of hydrocarbon. But it is not economic because commercially it is not produced.

2.3. Manufacturing Process of Hydrogen (H_2) gas

For synthesis of NH_3 both the gases namely, N_2 and H_2 should be as pure as possible. The N_2 gas comes from air, while H_2 is obtained from the different feedstock materials as discussed earlier. The processes for manufacturing of H_2 gas from different feedstock are:

(A) Steam reforming

(B) Non-catalytic partial oxidation (NCPO)

(C) Auto-thermal (combination of both A and B)

(D) Coal gasification

The major processes for manufacturing of H_2 gas are steam reforming and partial oxidation. Steam reforming process is followed in most of the NH_3 plant because of low cost of investment, operation and maintenance of plant. Both the processes produce a gas rich in H_2 and carbon monoxide (CO). The CO is then converted to carbon dioxide (CO_2) generating H_2 by reduction with steam in the shift converter.

2.3.1. Steam reforming

Steam reforming process is applicable only in case of natural gas and naptha. These two feedstocks have hydrocarbon content of more than 80%. The major constituent present in natural gas is methane (CH_4). Therefore, the reaction between steam (H_2O) and natural gas (with CH_4) to produce H_2 can be represented as:

$$\underset{\text{(Hydro carbon)}}{CH_4} + \underset{\text{(Steam)}}{H_2O} \xrightarrow[\text{Ni, Ca(AlO}_2)_2\text{, K}_2\text{O; Pressure 30-40 kg cm}^{-2}]{950\text{-}1400\ °C} CO + 3\,H_2 - \Delta H \text{ (Endothermic reaction)} \quad ...(2.14)$$

Similarly,

$$\underset{\text{(Hydro carbon)}}{CH_4} + \underset{\text{(Steam)}}{2\,H_2O} \xrightarrow[\text{Ni, Ca(AlO}_2)_2\text{, K}_2\text{O; Pressure 30-40 kg cm}^{-2}]{950\text{-}1400\ °C} CO_2 + 4\,H_2 - \Delta H \text{ (Endothermic reaction)} \quad ...(2.15)$$

Since in the above process, one mole of hydrocarbon (CH_4) is used to produce three moles of H_2, the overall H_2 content is increased by volume basis. Thus, the above process is called reforming process.

To optimize the production of H_2 gas, the important conditions or requirements to be maintained in steam reforming process are:

(a) ***Temperature***: Firstly, as per the theory of chemical equilibrium (Le Chatelier Principle, 1884), as this reaction is endothermic, it favours with increase in temperature of the reaction. Thus, temperature must be maintained at 950-1400 °C.

(b) ***Pressure***: Secondly, as the volume of the gas (product) increases, the decrease in pressure will help to increase the reaction in forward direction, while increase in pressure will reduce the forward direction reaction. To optimize the reaction, the pressure should be maintained at about 30-40 kg cm^{-2}.

(c) ***Catalyst***: Thirdly, the reaction is favoured when a suitable catalyst is used. Thus, in order to continue the forward reaction, pre-heated calcium aluminate [$Ca(AlO_2)_2$] is used as catalyst, while K_2O is used as promoter.

The steam reforming process of natural gas is usually carried out in two stages namely, (i) primary reforming, and (ii) secondary reforming. In the primary reforming stage, since the temperature goes down during the reaction, usual catalyst efficiency also goes down and only primary reforming takes place where about 85-95% of CH_4 is converted to H_2. When nickel (Ni) is used as the catalyst, it reacts with CO and forms nickel carbonyl compound [$Ni(CO)_4$]. Thus, the efficiency of reaction goes down.

$$Ni + 4\,CO \rightarrow Ni(CO)_4 \qquad \text{... (2.16)}$$

Secondary reforming is done to convert the unreacted or unconverted CH_4 (about 5-15% left after primary reforming) to H_2 gas. Thus, efficiency of the production of H_2 is increased. At this stage, a controlled amount of air is added to the process gas stream to provide sufficient O_2 to reform residual CH_4. Addition of air at this stage, not only increase the production of H_2 but also will introduce some amount of N_2 (through air) to maintain a ratio of N_2:H_2 at 1:3 in the NH_3 synthesis. Thus, purposes of introducing air at this stage are to increase the H_2 production, to increase the N_2 gas and to maintain the temperature (heat) during the reaction. The usual compositions of the gases leaving after secondary reforming stage by volume basis are 56% H_2, 12% CO, 8% CO_2 and 24% N_2.

According to Le Chatelier Principle, the synthesis of NH_3 should be done in absence of other gases or impurities *i.e.* N_2 and H_2 gases should be free from other impurities. So, removal of other gases from H_2 is must before synthesis of NH_3. Sulphur compounds may be present in the natural gas which must be removed to prevent poisoning of the reformer catalyst. Usually, the natural gas is passed through beds of iron oxide (Fe_2O_3) to remove hydrogen sulphide (H_2S) and then through activated carbon to remove S compounds.

2.3.2. Non-Catalytic Partial Oxidation (NCPO)

In non-catalytic partial oxidation (NCPO), the feed materials to the gas generator are steam, O_2 and hydrocarbon feedstock. The reaction involved in NCPO process may be represented as:

$$CH_4 + O \xrightarrow{1100\ ^\circ C} CO + 2H_2 \quad + \ \Delta H \text{ (Exothermic reaction)} \qquad \text{...(2.17)}$$

In NCPO process, the reaction is not complete. It is a partial oxidation. By complete oxidation, it will produce CO_2 and H_2O as per the reaction given below:

$$CH_4 + 2\,O_2 \rightarrow CO_2 + 2\,H_2O \text{ (complete oxidation)} \quad ...(2.18)$$

However, formation of H_2O is not desirable and only H_2 production is needed.

Advantages of NCPO over steam reforming are:

(i) No catalyst is required.

(ii) It requires low heat (it is an exothermic reaction).

(iii) It can tolerate more impurities present in hydrocarbon.

(iv) It can be used to a wide variety of feedstock ranging from natural gas to heavy fuel oil and crude petroleum.

Disadvantages of NCPO process is that an air-separation plant is required to provide O_2. Thus, initial investment cost is more.

The differences between steam reforming and NCPO processes are presented below:

Steam reforming	NCPO
• Reaction takes place with steam.	• Reactions takes place with O_2.
• It is an endothermic reaction.	• It is exothermic reaction.
• Here catalyst is used.	• No catalyst is used.
• Reaction:	• Reaction:
$CH_4 + \underset{\text{(Steam)}}{H_2O} \xrightarrow[\text{Ni catalyst, 30-400 kg cm}^{-2}]{950\text{-}1400\ ^\circ C} CO + 3\,H_2$	$CH_4 + O \xrightarrow{1100\ ^\circ C} CO + 2\,H_2$

2.3.3. Auto-thermal (combination of both A and B)

Auto-thermal process is a combination of steam reforming and partial oxidation. The feedstock ranges from natural gas and naptha to refinery gases and liquid petroleum gas. The feed materials for reactor are steam, air, O_2 and hydrocarbon feedstock. Special Ni-catalyst is present in the reactor.

2.3.4. Coal gasification

Coal can be used as feedstock for the production of H_2. This process is known as coal gasification. In India, coal is using as feedstock in three plants namely, Korba (MP), Ramagundam (AP) and Talcher (Orissa). The feed materials are coal or coke (C), steam and O_2. This process is also known as water gas route. The mixture of $CO+H_2$ is called water gas. It is similar to electrolysis of water to produce H_2. In case of electrolysis of water for breaking water molecule to H_2 and O_2 requires electricity. Whereas, in case of water gas route, no electricity is required. The heat energy comes from coke or coal or lignite as given below:

$$\underset{\text{(Red hot coke)}}{C} + \underset{\text{(Steam)}}{H_2O} \xrightarrow{1100\text{-}1400\ ^\circ C} \underset{\text{(Water gas)}}{CO + H_2} \quad \dots (2.19)$$

Coke is pulverized and used. Pulverization is done to increase the surface area of feed material to come in contact with steam. There are two coal gasification processes. These are (i) Lurgi process where mixture of steam and O_2 is passed through a bed of pulverized coal at 985 °C; and (ii) Koppers-Totzek process where a mixture of O_2 and coal-dust (pulverized coal) is passed into a gas generator and steam is introduced at 1100-1400 °C.

If air contains N_2 and O_2, then producer gas (N_2 + 2 CO) is formed.

$$\underset{(N_2 + O_2)}{\text{Air}} + 2\,C \rightarrow \underset{\text{(Producer gas)}}{N_2 + 2\,CO} \quad \dots (2.20)$$

If, coke is heated in presence of S-compound then carbonyl sulphide will be produced which is poisonous. Thus, it should be removed from the reactor.

2.4. Removal of CO from H_2

In all the above processes, carbon monoxide (CO) is also produced along with production of H_2. The H_2 gas should be as pure as possible before the reaction of H_2 and N_2 for NH_3 synthesis. Thus, CO produced during the production of H_2 must be removed.

Carbon monoxide is an inert gas, not soluble in water (insoluble) and other solvents. Thus, physical separation is not possible by simple absorption. Therefore, it is removed indirectly by converting it to some other form. There are two methods to remove CO from the system:

(a) ***Use of ammoniacal copper solution:*** The CO can be absorbed in ammoniacal copper solution and subsequently removed from the H_2 gas. But it is very costly and commercially not used.

(b) ***By shift conversion:*** The CO can be converted to CO_2 by oxidizing CO with steam or O_2. But steam is cheaper than O_2. The conversion of CO to CO_2 with steam is called shift conversion.

$$\underset{\text{(Steam)}}{CO + H_2O} \xrightarrow[Fe_2O_3,\ Cr_2O_3,\ Al_2O_3,\ CuO]{350\text{-}450\ °C} CO_2 + H_2 - \Delta E \text{ (Endothermic)} \quad \text{... (2.21)}$$

Generally, after production of H_2 gas and CO from different feedstocks, the gas mixture is passed through a converter containing a catalyst where it is mixed with steam and CO is converted to CO_2 as shown in the above reaction. It is important that during shift conversion some additional amount of H_2 is produced. Therefore, for the economic reason, conversion of CO to H_2 should be as high as possible.

The important factors affecting conversion CO to CO_2 are:

(i) Concentration of CO_2, CO and steam as the gases enter the shift conversion.

(ii) For conversion of CO to CO_2, a suitable catalyst is required. A mixture of Fe_2O_3 and Cr_2O_3 are usually used as catalyst. The catalyst mixture consisting of oxides of Al_2O_3, Cr_2O_3, CuO and Fe_2O_3 is also used. Recently, a catalyst composed of Zn, Cr and Cu oxides is used which is active at 175-250 °C.

(iii) Since, steam is converting CO to CO_2, it is an endothermic reaction. Thus, two gases must be pre-heated before the actual reaction. The catalyst is not active if temperature is below 350 °C. Thus, temperature is maintained at greater than 350 °C, preferably at 350-450 °C. After the reaction taken place, the reaction temperature goes down to 250 °C.

2.5. Removal of CO_2

The CO_2 thus formed by shift conversion is then removed from the reactor. Since CO_2 is soluble in water, sparingly soluble in most of the solvents, soluble in basic solvents (inorganic or organic) and least soluble in acid as it is also acidic in nature. Thus, CO_2 can be removed from the system by dissolving it with different solvents.

2.5.1. Removal of CO_2 by water

Most cheapest source is the water. Thus, in the purification step of most of the old NH_3 synthesis plant, CO_2 is removed by using water as the scrubbing agent. Water is passed through the gases at a pressure of 18 kg cm^{-2} and the CO_2 concentration is reduced to 1%. There are three important factors that should be considered while considering water as the reagent for CO_2 removal.

(a) Firstly, the efficiency of absorption of CO_2 by water is relatively less. Large amount of electricity is used and therefore the power cost is very high for pumping the large volume of water to remove CO_2.

(b) Secondly, H_2 absorbed by water is lost unless a recovery system is installed. Moreover, some amount of H_2 is lost during the process.

(c) Finally, CO_2 evolved from the water is not suitable for use in the manufacture of urea because of presence of H_2 and other inert gases.

After the removal of CO_2 from the H_2 gas, it should be regenerated. The regenerated CO_2 should then be used in the production of urea. For this purpose, the purity of CO_2 regenerated should be 99%. When water is used, it may be converted to form carbonic acid (H_2CO_3) and subsequently CO_2 (> 99% pure) may be obtained by boiling.

$$CO_2 + H_2O \rightarrow \underset{\text{(Carbonic acid)}}{H_2CO_3} \xrightarrow{\text{Boiling}} CO_2 + H_2O \quad ...(2.22)$$

The drawback of this process is that although CO_2 is regenerated, it will not be possible to get pure CO_2 as both H_2 and CO_2 are soluble in water.

2.5.2. Removal of CO_2 by hot potassium carbonate (K_2CO_3)

This process utilizes an aqueous solution of hot K_2CO_3 as the absorbent.

$$K_2CO_3 + CO_2 + H_2O \xrightarrow{85\ °C} 2\ KHCO_3 \quad ...(2.23)$$

Here, the mixture of CO_2 and H_2 is passed through a hot K_2CO_3 solution where most of the CO_2 is absorbed. More than 99% of CO_2 is removed by this process. In regeneration, $KHCO_3$ can be ignited to get pure CO_2.

$$2\ KHCO_3 \xrightarrow[110\ °C]{\text{ignited}} K_2CO_3 + H_2O + CO_2\ \text{(Pure)} \quad ...(2.24)$$

The above process has been modified by using a mixture of K_2CO_3 and Na_2CO_3 with arsenic oxide and other inorganic additives to effect more efficient removal of CO_2. This method is comparable with some other method like mono-ethanol amine (MEA). A solution of 15% Na_2CO_3 can also be used.

$$Na_2CO_3 + CO_2 + H_2O \xrightarrow{85\ °C} 2\ NaHCO_3 \quad ...(2.25)$$

2.5.3. Removal of CO_2 by using organic solvents like MEA or TEA method

The CO_2 can be removed by organic solvents like monoethanol amine (MEA). It is the most effective method for removing CO_2 from the synthesis gas. A solution of 15-30% MEA is usually used for this purpose. Besides MEA, there are some other organic solvents which can be used for removal of CO_2 such as tri-ethanol amine (TEA), methanol (CH_3OH) and propylene carbonate. Here, physical adsorption takes place between CO_2 and organic solvents. Methanol (CH_3OH) is used to remove CO_2 by absorbing CO_2 when the gas mixture is passed through methanol. But the disadvantage of use of methanol is that its boiling point is high and it may react with H_2S. Thus, other solvents are preferred. The MEA and TEA are basic in character and they have the properties to absorb CO_2 with the formation of corresponding salts. The purity of CO_2 released from MEA and TEA exceeds 99%. The disadvantages of these solvents are that it is corrosive (because of basic in nature) and relatively large amount of heat is required for regeneration of solution.

2.5.4. Fluor solvent

Some fluor solvents can be used for CO_2 removal. These are propylene carbonate, glycerol tri-acetate.

2.6. Final Gas Purification (Micropurification Stage)

After CO_2 removal, the final gas should be as pure as possible. The CO_2 concentration should be <10 ppm. Then, it is suitable for NH_3 synthesis. This can be done by (a) methanation, (b) copper solution, and (c) liquid N_2-wash.

(a) **Methanation:** This is known as catalytic methanation. It is just reverse to gas generation process of hydrocarbon. The methanation process involves heating the gases to about 300 °C and passing the hot gas through a nickel (Ni) based catalyst. Under such conditions, the CO_2 react with H_2 to form CH_4 and water.

$$CO + 3\,H_2 \underset{Ni}{\overset{300\text{-}350\,°C}{\rightleftharpoons}} CH_4 + H_2O \quad \dots (2.26)$$

$$CO_2 + 4\,H_2 \rightleftharpoons CH_4 + 2\,H_2O \quad \dots (2.27)$$

Water is removed from the gas, but methane is not removed because it does not harm the NH_3 synthesis. The advantages of methanation are (i) low equipment cost, and (ii) no operating cost except initial charge of catalyst. While, the main disadvantage of this method is the loss of a fraction of H_2 by reacting it with CO or CO_2.

(b) Copper solution: Copper solution is the oldest process used for final gas purification. It is a mixture solution of cuprous and cupric ammonium acetate, or cuprous and cupric ammonium formate or a mixture of above two. The final gas is passed through Cu-solution to reduce the CO_2 content to less than 5 ppm.

(c) Liquid N_2 wash: Liquid N_2 wash is used not only to remove the CO_2, but also to remove argon (Ar) and methane (CH_4). Thus, the final gas contains very little amounts of impurities. However, this process requires additional equipment for purification.

2.7. Synthesis of NH_3 (Fritz Haber and Carl Bosch, 1913)

Ammonia is synthesized by the reaction between H_2 and N_2 at elevated pressure and temperature in the presence of a catalyst.

$$N_2(g) + 3H_2(g) \underset{\substack{\text{Pressure 200-1000 kg cm}^{-2} \\ \text{Fe-oxides, Al, K, Mg, Ca-oxides}}}{\overset{400\text{-}450\ ^\circ\text{C}}{\rightleftharpoons}} 2NH_3(g) + \underset{\text{(heat)}}{\Delta E} \ (\text{Exothermic}, \Delta H = -92 \text{kJ mol}^{-1}) \quad ...(2.28)$$

The equilibrium reaction is governed by:

(a) Law of mass action – The rate of any reaction is proportional to the activity of the individual (Guldberg and Waage)

(b) Le Chatelier's Principle – Every system tries to attain in equilibrium. If the system is changed by changing its pressure, temperature and volume, it will shift to attain its new equilibrium or it will nullify the extra charge.

According to Le Chatelier's Principle, the production of NH_3 depends on the following conditions:

(i) The reaction should be done at low temperature since the reaction is exothermic. However, a temperature range from 400-450 °C is optimum in order to get as much NH_3 as possible.

(ii) Here, the volume is decreased as there are 4 molecules on the left hand side of the equation, but only 2 on the right. So the reaction is favoured by high pressure (200-1000 kg cm^{-2}).

(iii) Molecular ratio of N_2:H_2 should be 1:3 (stoichiometric ratio).

(iv) Fe-oxides is used as catalyst and oxides of Al, K, Mg and Ca (Al_2O_3, K_2O, MgO and CaO) are used as promoters. In the original Haber–Bosch (1913) method, they used osmium (Os) as the catalyst. However, osmium is very easily poisoned and requires frequent purification. In order to avoid such difficulties, the catalyst used universally is composed of Fe-oxides and promoters like Al, K, Mg and Ca-oxides. The catalyst is not active until the Fe-oxides have been reduced.

(v) Purity of gases should be as pure as possible. About 80% of the cost goes for the purification of the gases.

2.8. Anhydrous Ammonia (NH_3)

Anhydrous ammonia (NH_3) is the basic building block for almost all of the chemically derived nitrogenous fertilizers. Most of the NH_3 in the world is produced synthetically by reacting N_2 and H_2 gases although some of it is still recovered as a byproduct of coke or coal. Anhydrous ammonia has been gaining popularity as a nitrogenous fertilizer in the world, particularly in USA. Since, no further processing is required, NH_3 is the cheapest nitrogenous fertilizer for direct application. It is used to produce other fertilizers or is applied directly to the soil.

The ammonia (NH_3) is the key material for all most all types of fertilizers – not only nitrogenous, but also major phosphatic fertilizers or complex fertilizers as shown in the **Figure 2.1.**

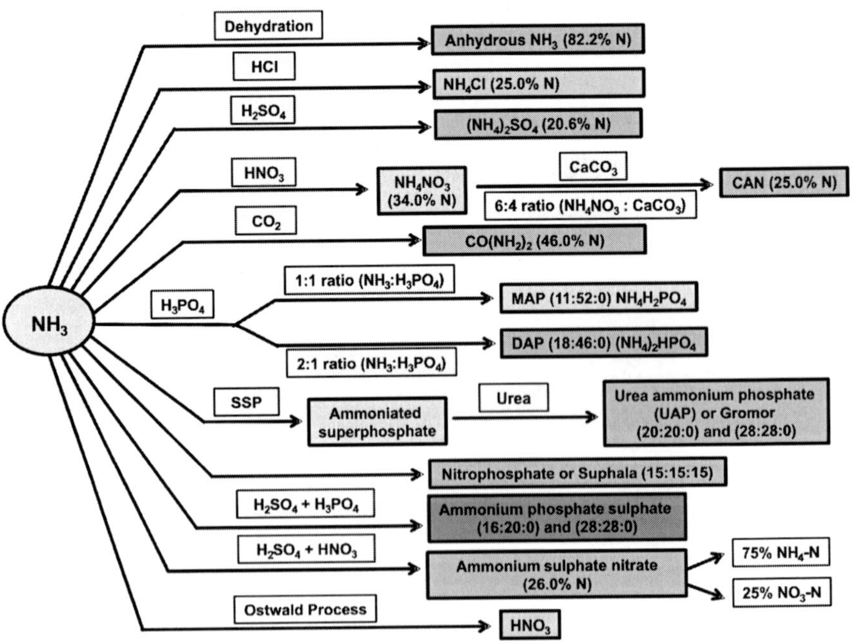

Figure 2.1. Schematic diagram sowing NH_3 as the key component for production of almost all the fertilizers

2.8.1. Properties of NH_3

(1) It contains 82.2% N, highest among all N-fertilizers.

(2) It resembles water in its behavior since they both have solid, liquid and gaseous state.

(3) Anhydrous NH_3 is lighter than air.

(4) It is a gas at ordinary condition and forms white cloud when NH_3 is released to atmosphere. This cloud is formed by the condensation of water vapour in the air surrounding the liquid NH_3 as it vaporizes.

(5) Under normal atmospheric pressure, anhydrous NH_3 in an open vessel boils constantly and escapes into atmosphere. To prevent this loss and conserve NH_3, it is generally stored in a high pressure tanks or vessels designed to withstand pressure 19 kg cm^{-2}.

(6) Ammonia gas liquefies at -33.3 °C. Gaseous form of NH_3 cannot be stored or transported and liquification is must. Thus, special type of refrigeration arrangements (-33.3 °C) is needed for storage and high pressure tanks or metering devices is required for application. For proper storage and application, important points that must be considered are (i) it needs special designed tanker or vessel; (ii) special steel vessel; and (iii) special application device (applicator).

(7) Vapour pressure of NH_3 increases with increase in temperature. This creates the problems of storage of NH_3. When temperature goes from 0 to 40 °C, vapour pressure goes up from 5 to 18.6 kg cm^{-2}. Common steel or iron tank cannot withstand this extra vapour pressure. Thus, special designed tank or vessel is used.

(8) Ammonia is alkaline in nature. Thus, applicator or metering device should be resistant to corrosive so that NH_3 does not affect the pipeline.

2.8.2. Application of anhydrous NH_3

Because it is a gas at atmospheric pressure, some anhydrous NH_3 may be lost to the atmosphere during and after application due to volatilization. About 10-15% of NH_3 will be lost by volatilization depending upon the soil and climatic condition. Factors associated with NH_3 losses are:

(i) Physical condition of the soil during application.

(ii) If, temperature is more, the loss of NH_3 will be more.

(iii) Coarse texture soils are more prone to losses than fine texture soil.

(iv) When soil moisture is less or if soil is dry then NH_3 losses is more.

(v) In order to reduce the losses, it should be injected and placed below 5-10 cm depth and covered with soil.

(vi) Equipment utilized for direct application of NH_3 consists of (i) a nurse tank, (b) an ammonia applicator, (c) an ammonia transfer system, and (iv) a tractor.

2.8.3. Soil conditions after application of NH_3

Immediately after application (injection) of anhydrous NH_3 into soil, it moves laterally as well as vertically (limited to < 5 cm), and a localized zone known as NH_3 retention zone is formed. A number of temporary yet dramatic changes occur in the NH_3 retention zone. These are:

(i) A retention zone of both ammonia (NH_3) and ammonium (NH_4^+) having circular to oval shape (3-13 cm diameter) is formed.

(ii) The concentration of both NH_3 and NH_4^+ increased in the range of 1000-3000 ppm.

(iii) The pH increased up to 9.0-9.5 in the vicinity of the point of application.

(iv) Concentration of NO_2^- (nitrite) increased to toxic levels (100-300 ppm). It happens due to the fact that *Nitrobacter* is much more sensitive to high pH than *Nitrosomonas*. Thus, the conversion of NO_2^- to NO_3^- is delayed and NO_2^- accumulates.

(v) Population of microorganisms decreased tremendously.

(vi) It is observed that free NH_3 (non-ionized NH_4) is more toxic to plants, animals, microorganisms. It can readily penetrate into cell membranes, while these tissues are relatively impermeable to NH_4^+.

2.8.4. Anhydrous NH_3 can also be applied through irrigation

Application of nutrients through irrigation water is convenient and less costly. Moreover, rate of application can be regulated according to the need of the crops. One of the major difficulties in applying NH_3 through irrigation is the loss of NH_3. It may be noted that losses of NH_3 is more at higher temperature, particularly at slow rate of release. Ammonia applied with irrigation is distributed with water. Thus, accurate application of irrigation is important. The concentration of NH_3 in irrigation water should not be more than 100 ppm otherwise losses will be more.

2.8.5. Mechanisms of NH_3 retention

2.8.5.1. Chemical mechanism of NH_3 retention

(1) ***Retention of NH_3 with H^+ ion:*** Ammonia applied as a fertilizer is retained by soil. However, it is not available to the plant and not even adsorbed or retained in soil unless it is converted to NH_4^+ form. For the formation of NH_4^+ ion, the reaction of NH_3 with proton (H^+) is necessary as shown below:

$$NH_3 + H^+ \rightarrow NH_4^+ \qquad ...(2.29)$$

The main source of proton is water. Presence of water in the field is very much essential. Thus, anhydrous NH_3 must be applied in moist soil, but not in dry condition where it will be lost as NH_3 volatilization. If water is not present, then it has to be placed at an appropriate depth or it has to be applied through NH_3-solution. Besides water, H^+ ion comes from following sources:

(i) Carboxyl group ($-COOH \rightarrow -COO^- + H^+$)

(ii) $Al^{3+} \rightarrow H^+$ (Hydrolysis of Al^{3+} generates H^+ ion)

(iii) Organic matter

(iv) Broken edges of clay minerals

(v) The proton can be provided by exchangeable H^+ and Al^{3+} ions in acid soils by the following reactions.

$$H^+ \text{(Clay)} + NH_3 \rightarrow NH_4^+ \text{(Clay)} \qquad ...(2.30)$$

$$Al(H_2O)(OH_2)H^+\text{(Clay)} + NH_3 \rightarrow NH_4^+ + [Al(H_2O)_5OH]^{2+} \qquad ...(2.31)$$

(2) Another mechanism involving adsorption of NH_3 by the exchangeable cation is analogous to the hydration of water. The NH_3 molecule like H_2O molecule has a pair of unshared electron on the N atom.

$$NH_3 \rightarrow \begin{matrix} & H \\ H: & \ddot{\underset{\cdot\cdot}{N}} \\ & H \end{matrix} \text{(Lewis base)}$$

The two molecules have been shown to compete with each other.

(3) Reaction of NH_3 with hydroxyl group (-OH) of specific type of clay minerals. The reaction can be represented by the following equation:

$$NH_3 + HOX \rightarrow NH_4OX \ (X = \text{clay})$$

(4) Ammonia also reacts with organic colloids. About 50% NH_3 retention in soil is due to organic matter.

(5) Precipitation of Ca^{2+} and Mg^{2+} as carbonates in the presence of CO_2 and increasing availability of the exchange sites for reaction with NH_4^+.

2.8.5.2. Physical mechanism of NH_3 retention

(1) Entrapment of NH_3 between the lattices of smectite type clay minerals.

(2) Replacement of interlayer cations in an expanded clay lattice by NH_4^+.

(3) Adsorption by clay minerals and organic compound by H-bonding.

The relative mechanisms vary from soil to soil and also be influenced by environmental conditions.

2.8.6. Factors affecting NH_3 retention in soil

(1) ***Soil moisture:*** Retention of NH_3 in soil is maximum at or near field capacity moisture level. While both dry soils as well as soils having moisture content more than field capacity decreases NH_3 retention.

(2) ***Soil texture:*** Retention of NH_3 increases with increase in clay content. Thus, NH_3 retention is more in fine texture soil than coarse texture soil.

(3) ***Application depth:*** Retention of NH_3 increases when it is placed at soil depth of 5 cm in case of silt loam and 10 cm in case of fine sandy loam soil.

(4) ***Cation exchange capacity (CEC):*** The retention of NH_3 increases with CEC of soil.

(5) ***Organic matter:*** More is the soil organic matter (SOM) content, more is the NH_3 retention. At least 50% of the NH_3 retention is attributed to SOM.

(6) ***pH:*** The NH_3 loss increases with increase in pH, especially when fertilizers are applied on the surface.

2.8.7. Effect of NH_3 on soil properties

The drastic modifications within the soil zone as a result of application of high concentration of ammonia, aqua ammonia, urea and other ammoniacal fertilizers exert considerable influences on various soil properties, and ultimately on the crop responses. These are:

(1) ***Effect on soil microorganisms and nitrification:*** High concentration of NH_3 and NH_4^+ in combination with basic soil pH and high osmotic potential

results in a partial and temporary sterilization of soil within the retention zone. Bacterial activity is probably affected mostly by free NH_3 while fungi are depressed by high pH. Nitrification is restricted due to addition of NH_4^+-forming and NH_4^+ containing fertilizers.

(2) ***Effect on germination and seedling growth:*** High concentration of NH_3 (>1000 ppm NH_3 is harmful), NH_4^+ and NO_2^- affects the germination of seed and seedlings growth.

(3) ***Effect on SOM:*** The NH_4OH produced by the reaction of alkaline hydrolyzing fertilizers in soil will dissolve and hydrolyze certain fractions of SOM. However, these effects on SOM are considered temporary and disappear.

(4) ***Effect on nutrient availability:*** Solubilization of SOM is expected temporarily to increase the availability of nutrients associated with the organic fractions of soils. Readily available P increases on application of anhydrous NH_3.

2.8.8. Application problem of anhydrous NH_3

There are several problems associated with application of anhydrous NH_3.

(1) Anhydrous NH_3 is having pungent smell.

(2) It has physiological toxicity. Concentration >200 ppm NH_3 in air and if inhaled for 2 h is toxic to human beings. So, a special apparatus like hand gloves, mask and glasses should be used to handle anhydrous NH_3.

2.9. UREA [(NH_2-CO-NH_2) or $CO(NH_2)_2$]

History: Urea is the most widely used nitrogenous fertilizer in the world today including India. Urea is the first organic compound synthesized from inorganic compounds. It was first separated from urine by **Rouelle (1773)**, hence the name urea. The discovery of urea was an accident event. **Friedrich Wöhler (1828)** first synthesized urea by the isomerization of ammonium cyanate (NH_4CNO). This synthesis is of great historical significance because for the first time an organic compound was produced from the inorganic compound. Earlier it was thought that organic compound cannot be produced from inorganic compound. But, urea is produced from organic compound. This gave new light in fertilizer chemistry. Since, then numerous processes have been developed for urea synthesis. It is considered to be the starting point of modern organic chemistry. Although the Wöhler reaction deals with the conversion of ammonium cyanate, this salt appears only as an (unstable) intermediate compound. Wöhler demonstrated the reaction as a combination of cyanic acid (HNCO) and ammonia (NH_3). During heating of NH_4CNO, rearrangement of atoms takes place because of isomerization, resulting in the formation of urea as shown in equation below.

$$\underset{\text{(Ammonium cyanate)}}{NH_4CNO} \xrightarrow{\text{heating}} \underset{\text{(Urea)}}{NH_2\text{-}CO\text{-}NH_2} \quad ...(2.32)$$

Alternately, the reaction can be carried out with lead cyanate and ammonia to form ammonium cyanate. Ammonium cyanate decomposes to ammonia and cyanic acid which in turn reacts to produce urea by isomerization or tautomerization:

$$Pb(NCO)_2 + 2\ NH_3 + 2\ H_2O \rightarrow Pb(OH)_2 + 2\ NH_4NCO \quad ...(2.33)$$

$$NH_4NCO \rightarrow NH_3 + HNCO \rightarrow (NH_2\text{-}CO\text{-}NH_2) \quad ...(2.34)$$

The **synthesis of urea** from cyanic acid and ammonia through the intermediate of ammonium cyanate is known as the **Wöhler synthesis**.

*Note: *The difference between organic and inorganic compound is that in organic compound, presence of 'C' is must, but it is not the reverse. That means all organic compound must have 'C', but all compounds containing 'C' is not organic compound. Thus, carbonates (CO_3^{2-}) are not organic compound.*

Urea is a derivative of carbonic acid (H_2CO_3 or HO–CO–OH). When H_2CO_3 is reacted with NH_3, then stepwise replacement of –OH group in H_2CO_3 will occur by $-NH_2$ group, as a result of this conversion, it will produce urea.

$$\underset{\text{(Carbonic acid)}}{HO\text{–}CO\text{–}OH} \xrightarrow[-H_2O]{+NH_3} \underset{\text{(Mono-amidal carbonic acid)}}{HO\text{–}CO\text{–}NH_2} \xrightarrow[-H_2O]{+NH_3} \underset{\text{(Di-amidal carbonic acid) (Urea)}}{NH_2\text{–}CO\text{–}NH_2} \quad ...(2.35)$$

In India, the first urea plant was installed at Sindri in Bihar in 1959. Presently there are 31 urea plants producing urea in the country, 28 of them are using natural gas and three are using naptha as the feedstock. About 94% of India's urea production capacity is based on natural gas, while only 6% is naptha-based.

2.9.1. Methods of Manufacture of Urea

Commercially, urea is produced by chemical reaction with anhydrous liquid NH_3 and gaseous CO_2 in a reactor maintained at temperature range of 180-200 °C and pressure 200-350 kg cm^{-2}. The chemical reaction takes place in two steps.

In the first step, ammonium carbamate is produced by reacting NH_3 and CO_2 which is an exothermic (37.7 kcal/g-mole) process. In the second step, ammonium carbamate is decomposed and urea is formed which is moderately endothermic (–7.7 kcal/g-mole).

$$2\,NH_3 + CO_2 \xrightarrow[200\text{-}350\ \text{kg cm}^{-2}]{180\text{-}200\ ^\circ C} \underset{\text{(Ammonium carbamate)}}{NH_4COONH_2} \qquad ...(2.36)$$

$$\underset{\text{(Ammonium carbamate)}}{NH_4COONH_2} \xrightarrow[\text{(Decomposition)}]{250\ ^\circ C} \underset{\text{(Urea)}}{NH_2\text{–}CO\text{–}NH_2} + H_2O \qquad ...(2.37)$$

The operating conditions to be maintained during the first step are: (a) temperature 180-200 °C; (b) pressure 200-350 kg cm^{-2}; (c) the stoichiometric ratio of CO_2:NH_3 is 1:2. However, in practice, the ratio of 1:4 and retention time of 20-30 min should be maintained to accelerate ammonium carbamate formation.

The reaction products containing mixture of urea, NH_3, CO_2, NH_4COONH_2 (ammonium carbamate) and H_2O leaves the reactor continuously through a pressure reducing valve. It was then passed through a separator where all the unconverted NH_3, CO_2 and NH_4COONH_2 are separated. The urea solution can be used as fertilizer or it can be further processed to produce solid urea in the form of crystals or prills or granules. The mixed unconverted NH_3 and CO_2 can be recovered and recycled to the reactor. The untreated NH_3 and CO_2 can also be used for the production of other fertilizers *e.g.* $(NH_4)_2SO_4$ or NH_4NO_3.

2.9.2. Routes of manufacture of urea

In the above reaction, only one-third (~33%) of NH_3 and two-third (~66%) of CO_2 passed in the reactor are converted to form urea. The remaining unconverted NH_3 (~66%) and CO_2 (~33%) are recycled back to reactor to convert them to form urea. Depending on the degree of recycling, the routes of manufacture of urea may be called as "**once-through**", "**partial-recycle**" or "**total-recycle**". Most new plants now use total-recycle process. However, once-through or partial-recycle processes have been popular in some countries and in some cases may still be preferred.

(a) In "**once-through**" process, NH_3 and CO_2 are passed through the reactor only once. It is the simplest and least expensive (in terms of both capital investment and operating costs) among the three processes. Whatever amount of urea is formed is collected. These gases cannot be used again and again to produce urea. The unconverted gases called "**off gases**" however, should be used for other purposes like for production of ammonium sulphate, liquid NH_3, ammonium chloride, ammonium bicarbonate, sodium bicarbonate, sodium carbonate, dry ice, *etc*. To utilize these gases, therefore, in most of the urea-plant, a co-plant is constructed along with the urea-plant.

(b) In case of **partial-recycle**, as the name implies, parts of the off-gases *i.e.* NH_3 and CO_2 are recycled back to the urea-reactor till about 85% of NH_3 gets converted to form urea. Even after partial recycle process, there should be a co-plant to utilize unreacted off-gases.

(c) In the **total-recycle process**, all the unconverted or unreacted gases are recycled back to the urea reactor several times so that more than 99% of the gases are utilized to form urea. In the total recycle, no co-plant is necessary. Thus, maximum efficiency is obtained through total recycle route. However, cost of investment and operation of urea-plant is more in total recycle route.

The urea solution (~80%) thus obtained either 'once through' or 'partial recycle' or 'total recycle' is then saturated → crystallized → prilled → dried → screened → coated → bagged → marketed.

2.9.3. Properties of urea

2.9.3.1. Physical Properties

(1) Urea is an organic compound containing 46% N, entirely in the organic form (amide). Pure grade urea contains 46.6% N.

(2) The critical relative humidity (CRH) signifies the capacity of fertilizer to absorb moisture from atmosphere. The CRH value of urea is 72 (Table 2.1). This indicates that urea is moderately hygroscopic. Thus, urea is free-flowing and does not form cake, lumps or aggregates.

(3) Urea is an acid-forming fertilizer (equivalent acidity is 80) but is less acidic than ammonium sulphate and ammonium chloride. The term 'equivalent acidity' indicates the amount of lime ($CaCO_3$) required for neutralizing the acidity caused by 100 kg of nitrogenous fertilizer. This signifies that by application of 100 kg urea, it produces acidity which needs 80 kg of $CaCO_3$ to neutralize it. Therefore, equivalent acidity of urea is 80. Similarly, the equivalent acidity of ammonium sulphate is 107, ammonium chloride is 127 and anhydrous ammonia is 148 (Table 2.1).

Table 2.1. Various nitrogenous fertilizers and their equivalent acidity and critical relative humidity

Nitrogenous fertilizers	Equivalent acidity	Critical relative humidity (CRH) at 25 °C
Urea (NH_2-CO-NH_2)	80	72
Ammonium sulphates [$(NH_4)_2SO_4$]	107	81
Ammonium chloride (NH_4Cl)	127	76
Ammonium nitrate (NH_4NO_3)	60	59
Calcium ammonium nitrate (CAN)	0	-
Anhydrous ammonia (NH_3)	148	liquid

(4) Specific gravity of urea is 1.45 g cm^{-3}.

(5) The melting point of urea is 132.7 °C.

(6) Formation of biuret (NH_2-CO-NH-CO-NH_2): When urea is heated at 132.7 °C, it decomposes and liberates ammonia (NH_3). The pungent smell of NH_3 in urea indicates that it is decomposing, resulting in the formation of biuret. Biuret is also formed when urea or concentrated urea solutions are exposed to elevated temperature (132.7 °C) during granulation process by melting and cooling or evaporation-prilling process of urea. Here, two moles of urea combines together to form biuret with the liberation of one mole of NH_3.

$$\underset{\text{(Urea)}}{NH_2\text{-}CO\text{-}N\ H_2} + \underset{\text{(Urea)}}{H_2N\text{-}CO\text{-}NH_2} \xrightarrow[-NH_3]{\text{Heating (132.7 °C)}} \underset{\text{(Biuret)}}{NH_2\text{-}CO\text{-}NH\text{-}CO\text{-}NH_2} \quad ...(2.38)$$

Biuret content in urea, if present in excess, is toxic to crops. As per the Fertiliser (Control) Order (FCO, 1985), the maximum permissible limit of biuret content in urea is 1.5% for soil application and 0.5% for foliar spray beyond which it is toxic to plants. Based on the biuret tolerance, the crops can be grouped into three categories *i.e.* (i) sensitive, (ii) semi-sensitive, and (iii) tolerant crops.

(i) ***Sensitive crops*** can tolerate up to **0.25% biuret**, *e.g.* vegetables, fruits (citrus) and horticultural crops.

(ii) ***Semi-sensitive crops*** can tolerate **1.5% biuret**, *e.g.* cereals crops.

(iii) ***Tolerant crops*** can tolerate **2.5% biuret**, *e.g.* fibre crops like jute, mesta.

(7) Solubility: Urea is highly soluble in water. Solubility of urea is one of the main physical constraints. It dissolve with absorption of heat or endothermic *i.e.* solution become cooler. Some heat is required for its dissolution. Solubility of urea is 45.5 g $100g^{-1}$ (*i.e.* 45.5 g urea may be dissolved in 100 mL of water). Because of high solubility, it gets lost due to leaching and run-off, when there is heavy rainfall or excessive irrigation after application of urea. Urea as such is mobile in soil and moves vertically as well as horizontally. Because of high solubility whole area molecule will be lost from soil through leaching and run-off. Thus, water management is very important when urea is applied in the field. This property of high solubility of urea can be manipulated by several ways *viz.*, changing size of particles and coating urea by some means.

(8) Generally, it is in crystalline form, but fertilizer grade urea marketed in India is in the form of prills, often called as prilled urea (size of 2-4 mm diameter). Urea super granule (USG) of 10-11 mm size is used in deep water rice cultivation. It is quite popular in Bangladesh.

(9) Urea is corrosive in nature, thus cause problem to lining of reactor. So, after some time lining should be replaced or some extra precaution should be taken by using steel lining or oil can be coated to prevent corrosion.

2.9.3.2. Chemical Properties of Urea

(1) ***Hydrolysis:*** After application of urea in soil, it undergoes hydrolysis where organic form of N is transformed to inorganic form of ammonium carbonate [$(NH_4)_2CO_3$]. It is a biochemical process, mediated by urease enzyme. The time required to complete hydrolysis in normal soil is 2-3 days, while it requires 7 days in alkaline soil. It is a two steps process, wherein urea is first hydrolyzed to form ammonium carbamate and then ammonium carbonate as follows:

$$\underset{}{NH_2\text{-}CO\text{-}NH_2 + H_2O} \xrightarrow[\text{Urease}]{\text{Hydrolysis}} \underset{\text{Ammonium carbamate}}{NH_4\text{-}COO\text{-}NH_2 + H_2O} \longrightarrow \underset{\text{Ammonium carbonate}}{(NH_4)_2CO_3} \quad ...(2.39)$$

The entire process is carried by urease enzyme. Sources of enzyme urease are plants, organic matter, roots exudates, soil organisms, decomposed cells, *etc*. Urease enzyme may be of two types namely, cellular and extra-cellular. These are protein substance. The activity of urease enzyme depends on (a) pH, (b) salt concentration, (c) organic matter, (d) clay content, (e) temperature, and (f) moisture.

(a) ***pH***: In neutral soil, the time required to complete hydrolysis is 48-72 h. However, in alkaline soil, it requires 7 days to complete hydrolysis. Thus, in neutral soil, hydrolysis is faster than alkaline soil. In normal soil, hydrolysis is fast, whereas in problem soils it is slow.

(b) ***Salt concentration***: The activity of urease is suppressed if some mercuric salt is added. Similarly, copper salt also reduces the activity.

(c) ***Temperature:*** It also influences the activity of urease enzyme. Since urease is a protein substance, at high temperature protein is denatured, thus activity of urease is reduced at high temperature.

(d) ***Organic matter***: More is the organic matter content in soil, less is the activity. If, organic matter content is increased by application of farmyard manure (FYM), green manure (GM), *etc.* then activity of urease is reduced *i.e.* adversely affected the hydrolysis process.

(e) ***Moisture content***: At field capacity moisture level, the urease activity is maximum. The activity of urease reduces under dry conditions as well as anaerobic and waterlogged conditions.

(f) ***Clay content***: High clay content in soil means urease activity is low. Thus, hydrolysis is more in sandy soil than clay soil. High clay content helps in adsorbing more urea molecule. This is because urea molecule forms positive charged ion due to ***keto-enol*** formation or zwitterion formation. Because of formation of positive charged urea molecules it is adsorbed by the negatively charged clay. Thus, more is the clay content more is the adsorption.

[*A **zwitterion** is originally derived from a German word "**zwitter**" which means **hybrid** or **hermaphrodite**. A zwitterion is normally described as an ion containing two **functional groups**. In simple terms, it is an ion possessing both **positive** and **negative electrical charge** at different locations within a molecule. Since the molecules contain two opposite charges, **they are electrically neutral or the net charge is zero**.*]

Ammonium carbonate is then transformed to NH_4^+ and CO_3^{2-} as shown in equation 2.40 or NH_3, CO_2 and H_2O as shown in equation 2.41.

$$(NH_4)_2CO_3 \rightarrow 2\ NH_4^+ + CO_3^{2-} \quad \text{... (2.40)}$$

or,

$$(NH_4)_2CO_3 \rightarrow 2\ NH_3 + CO_2 + H_2O \quad \text{... (2.41)}$$

- Fate of $(NH_4)_2CO_3$ indicates that a portion of NH_4^+ is taken by plants (mainly rice). How much is taken by plants depends on the genetic characteristics of plants *i.e.* types of crops and varieties.

- A major share of NH_4^+ is lost by volatilization as NH_3 (25-30%).
- A small amount of NH_4^+ is fixed by clay.
- Another portion of NH_4^+ is converted to form first NO_2^- and then NO_3^- by the process called nitrification, mediated by two groups of bacteria namely, *Nitrosomonas* and *Nitrobactor*, respectively.
- Part of NO_3^-, thus formed is utilized by plants. Part of it is subjected to loss due to leaching.
- Further, in waterlogged or puddled soil, part of NO_3^- goes to reduced layer where it is lost as oxides of N (NO, NO_2 and N_2O) by denitrification process.

2.9.3.3. Factors affecting NH_3 volatilization

The extent of **NH_3 volatilization** depends on several factors. These are:

(i) ***pH:*** The conversion of $(NH_4)_2CO_3$ to NH_3 is enhanced by high pH of soil. If soil pH >8.5, volatilization is high. In alkaline soil (pH >8.5), it is lost as NH_3 volatilization. In saline-alkali soils (pH > 8.5 and ESP >15%), the $(NH_4)_2CO_3$ is totally unstable and lost.

(ii) ***Temperature***: High temperature (>35 °C) leads to high volatilization.

(iii) ***Clay content***: High clay content in soil means, more absorption of NH_3, thus less volatilization.

(iv) ***Cation exchange capacity (CEC)***: If, CEC of the soil is low, then volatilization is high.

(v) ***Wind velocity***: High wind velocity resulted in high volatilization.

(vi) ***$CaCO_3$ content***: Higher is the $CaCO_3$ content, higher is the NH_3 volatilization.

(vii) ***Dry soil***: In dry soil, NH_3 volatilization is higher than the moist soil.

(viii) ***Rate of N application***: When the rate of release is equal to the rate of uptake by plant, then there will be no volatilization loss. But, if the rate of release is greater than the rate of uptake, there will be high volatilization loss. Higher is the rate of N application, higher is the rate of volatilization.

(ix) ***Method of application***: Top dressed urea resulted in greater NH_3 volatilization. While, placement of urea reduce the losses.

(x) ***Seasonal variation***: During summer, NH_3 volatilization is more than winter season.

(xi) ***Moisture***: Urea should be applied to moist soil and mixed thoroughly to get maximum use efficiency. It should not be applied in dry condition.

2.9.3.4. What will be the alternative method to apply urea so that urea loss is minimized?

(i) If, the rate of hydrolysis is reduced then the loss of NH_3 volatilization can be reduced or minimized. This can be done by using urease inhibitors. The urea molecule will remain in the soil for longer time. Although there will be some losses of urea as such by leaching.

(ii) By changing the method of application (say placement) loss of NH_3 volatilization can be reduced.

(iii) By reducing the solubility of urea using some coating materials or by increasing bigger particle size like urea super granule (USG).

2.9.4. Reactions of Urea with Acids

Urea can reacts with mineral acids like H_2SO_4, H_3PO_4 and HNO_3 to form different salts. Urea molecule has two NH_2 group and it reacts with mineral acids to form urea-sulphate, urea-phosphate and urea-nitrate, respectively.

$$NH_2CONH_2 + H_2SO_4 \rightarrow \underset{\text{(Urea-sulphate)}}{NH_2CONH_2.H_2SO_4} \qquad ...(2.42)$$

$$NH_2CONH_2 + H_3PO_4 \rightarrow \underset{\text{(Urea-phosphate)}}{NH_2CONH_2.H_3PO_4} \qquad ...(2.43)$$

$$NH_2CONH_2 + HNO_3 \rightarrow \underset{\text{(Urea-nitrate)}}{NH_2CONH_2.HNO_3} \qquad ...(2.44)$$

The product of **urea-sulphate** can be used in S-deficient soil. The physical condition of **urea-phosphate** is very good as it is less hydroscopic and it can used as source of both N and P. While, **urea-nitrate** is most hygroscopic and difficult in storage, transportation and application.

2.9.5. Reaction of Urea with Monocalcium Phosphate [$Ca(H_2PO_4)_2.H_2O$]

It is an important property as far as for storage point of view for different fertilizers. It is found that urea and single superphosphate (SSP) cannot be stored together for long period, but can be mixed and applied to soil just after mixing. Within 2-3 days of mixing of urea and SSP, reactions take place between urea (NH_2CONH_2) and monocalcium phosphate [MCP, $Ca(H_2PO_4)_2.H_2O$] present in SSP.

Here, crystallized H_2O present in MCP is liberated as free water and forms cake. The water soluble P in MCP is also converted to water insoluble, but citrate soluble dicalcium phosphate (DCP) as shown in equation 2.45.

$$Ca(H_2PO_4)_2.H_2O \rightarrow \underset{(DCP)}{CaHPO_4} + H_3PO_4 + H_2O \quad ...(2.45)$$
(MCP)

Further, urea reacts with H_3PO_4 liberated from MCP and form urea-phosphate as in equation 2.46.

$$NH_2CONH_2 + H_3PO_4 \rightarrow \underset{\text{(Urea-phosphate)}}{NH_2CONH_2.H_3PO_4} \quad ...(2.46)$$

The above reactions signify following three things:

(i) Firstly, it will reduce the availability of P present in SSP due to conversion of MCP in SSP to DCP.

(ii) Secondly, liberated free water forms cakes or aggregates, thereby physical properties is also deteriorated with the change of MCP.

(iii) Formation of cakes or aggregates enhanced problems in storage, transportation and application of urea and SSP.

Urea also reacts with MCP to form a product or cake or aggregate known as Adduct as shown in equations 2.47.

$$4\ NH_2CONH_2 + Ca(H_2PO_4).H_2O \rightleftharpoons 4\ NH_2CONH_2.Ca(H_2PO_4)_2 + H_2O \quad ...(2.47)$$

As per the specification mentioned in Fertiliser (Control) Order (FCO, 1985), free H_3PO_4 is allowed in SSP (3% H_3PO_4) and TSP (4% H_3PO_4). This free H_3PO_4 prevents the conversion of MCP to DCP. When H_3PO_4 is not present, then formation of DCP is very fast.

2.9.6. Reactions with Nitrous acid (Van Shyke reaction)

Under certain situations, urea reacts with nitrous acid (HNO_2) and form N_2 gas and lost. This process is known as Van Shyke reaction. About 3-5% of loss of N_2 occurs through this process.

$$NH_2CONH_2 + 2\ HNO_2 \rightarrow 2\ N_2 (\uparrow) + CO_2 (\uparrow) + 3\ H_2O \quad ...(2.48)$$

The above reaction occurs when the situations are favourable for accumulation of HNO_2. In nitrification process, NH_4^+ is converted first to NO_2^- and then NO_3^- as follows:

$$NH_4^+ \xrightarrow[\textit{Nitrosomonas}]{+O_2} NO_2^- \xrightarrow[\textit{Nitrobactor}]{+O_2} NO_3^- \quad ...(2.49)$$

(i) Under oxygen stress condition, nitrification is restricted to NO_2^- only because of reduced activity of *Nitrobactor*. Thus, NO_2^- concentration accumulates in the vicinity of urea molecule in the form of nitrous acid ($NO_3^- + H^+ \rightarrow HNO_2$), resulting conditions suitable for formation of Van Shyke reaction.

(ii) After 2 days of application of urea, pH is raised due to formation of $(NH_4)_2CO_3$. It is proportional to rate/level of N application. In high pH condition, *Nitrosomonas* is active, but *Nitrobactor*'s activity is less and NO_2^- accumulation takes place in soil.

(iii) If nitrification inhibitor is used then formation of NO_2^- is more; under such condition, urea is converted to elemental N_2.

Van Shyke reaction can be used for quantitative estimation of urea-N. This can be done by collecting the liberated N_2 under water where CO_2 is also adsorbed by water; while N_2 is collected in the gas cylinder. The amount of CO_2 is determined by titration with K_2CO_3.

2.9.7. Reaction with Sodium hypobromide (NaOBr)

Urea reacts with sodium hypobromide (NaOBr) to form elemental N_2 as shown below:

$$NH_2CONH_2 + 3\ NaOBr \rightarrow 3\ NaBr + CO_2\ (\uparrow) + N_2\ (\uparrow) + 2\ H_2O \quad \ldots(2.50)$$

This reaction is similar to that of reaction with HNO_2. Here, N_2 gas is released. The liberated N_2 can be collected and determined by standard procedure. For this, note down the volume of N_2, temperature and pressure. Then calculate in NTP (normal temperature and pressure) or STP (standard temperature and pressure) and estimate the total quantity of urea present. This method can be used for determination of urea-N.

2.9.8. Keto-Enol Tautomerism

Generally, it is believed that urea is not retained by any soil constituents. But, there is one hypothesis, where urea is retained in soil. This hypothesis is known as ***Keto-Enol*** tautomerism or formation of ***enol (OH⁻)*** group from ***keto (=CO)*** group present in urea. It is 1-2% of total urea applied. Soil colloid can retain this urea.

$$\underset{\text{(Keto-form)}}{H_2N{-}\overset{\overset{O}{||}}{C}{-}NH_2} \underset{\text{Tautomerism}}{\rightleftharpoons} \underset{\text{(Enol-form)}}{NH_2{-}\overset{\overset{OH^-}{|}}{C}{=}NH^+}$$

Here, keto group is changed to enol group. Due to this, urea has got polarity. Otherwise, it is a non-polar compound. Due to this changes, it is oriented to positively charged molecules which is fixed on the negatively charged clay. It is an electrostatic bond. This urea is not utilized by plant.

The transformation or fate of urea in soil upon application and its subsequent utilization by crop and losses of N is presented in Figure 2.2.

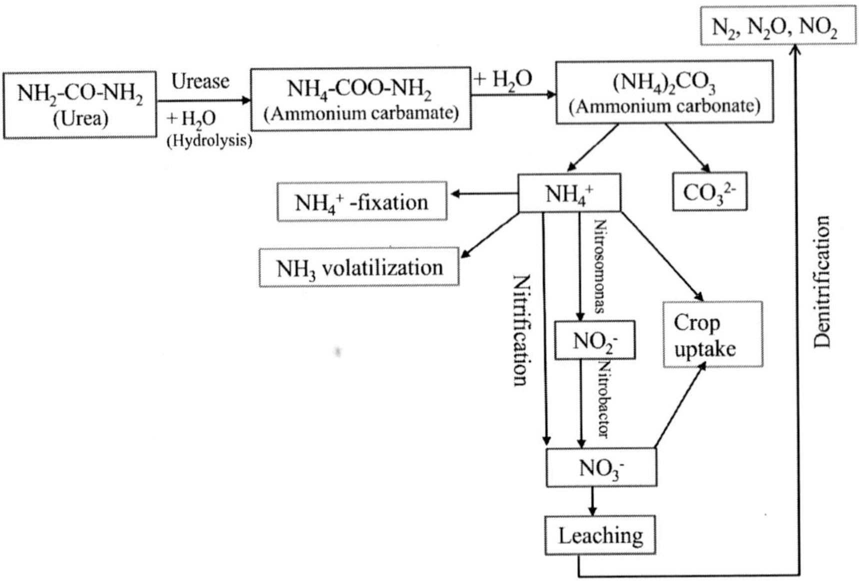

Figure 2.2. Transformation of urea in soil and uptake of NH_4^+ and NO_3^--N by plants

2.10. Ammonium Sulphate [$(NH_4)_2SO_4$]

2.10.1. Manufacturing Process (Chemistry of Production)

Ammonium sulphate [$(NH_4)_2SO_4$] can be commercially manufactured by following processes:

2.10.1.1. By-product of steel industry

Majority of the ammonium sulphate comes from by product of steel industries. It was first produced in India as a by-product of the steel industry in 1933 at Jamshedpur, Jharkhand. During burning of coke or coal in metallurgical industries, large amounts of coke-oven gases, rich in NH_3 are liberated through chimney. This coke oven gases are not allowed to escape to the environment because of possible environmental pollution. This liberated NH_3 gas is passed through dilute H_2SO_4 and $(NH_4)_2SO_4$ is produced as per equation 2.51.

$$2\,NH_3 + H_2SO_4 \rightarrow (NH_4)_2SO_4 \quad ...(2.51)$$

2.10.1.2. Direct reaction between anhydrous NH_3 and H_2SO_4

Ammonium sulphate can be prepared by direct reaction between anhydrous NH_3 and H_2SO_4. The process has not found much acceptance because of H_2SO_4 being costly as N-fertilizer compared to urea.

$$2\,NH_3 + H_2SO_4 \rightarrow (NH_4)_2SO_4 \qquad \text{...(2.52)}$$

2.10.1.3. Destructive distillation of coal

During the destructive distillation of coal (distillation in absence of O_2) NH_3 is produced in a closed reactor. In the destructive distillation, three distinct fractions namely, (i) gases, (ii) liquid and (iii) solid fractions are generated. Coal gases contain about 17% of coal used for distillation. While, liquid fraction is called NH_3 liquor and it contains 3-5% NH_3. The solid fraction is a dark coloured, viscous, thick phase, called tar or pitch. It is left at the bottom of the reactor.

The NH_3 thus produced can be absorbed in sulphuric acid to produce ammonium sulphate.

$$2\,NH_3 + H_2SO_4 \rightarrow (NH_4)_2SO_4 \qquad \text{...(2.53)}$$

2.10.1.4. Caprolactum Process

Caprolactum is an intermediate by-product of synthetic fibre (nylon polymer) industry. During synthesis of nylon polymer, large amount of NH_3 comes out. The H_2SO_4 is used as a catalyst and NH_3 is the product of hydrolysis. The liberated NH_3 is then treated with H_2SO_4 and ammonium sulphate is prepared.

$$2\,NH_3 + H_2SO_4 \rightarrow (NH_4)_2SO_4 \qquad \text{...(2.54)}$$

This process was adopted by Gujarat State Fertilizer Corporation (GSFC), Vadodara, Gujarat in 1974.

2.10.1.5. Gypsum Process

This is the most important process of manufacturig $(NH_4)_2SO_4$ commercially. In this process, ammonium sulpahte is prepared in two steps. In the first step, ammonium carbonate $[(NH_4)_2CO_3]$ is prepared by reacting NH_3 gas with CO_2 gas. Mixture of off-gases generated during production of urea (*i.e.* mixture of NH_3 and CO_2) is also used for preparation of $(NH_4)_2CO_3$

$$2\,NH_3 + CO_2 + H_2O \rightarrow (NH_4)_2CO_3 \qquad \text{...(2.55)}$$

In the second step, gypsum ($CaSO_4$) is reacted with a solution of ammonium carbonate to form ammonium sulphate. Calcium carbonate is separated from ammonium sulphate by filtration.

$$(NH_4)_2CO_3 + CaSO_4 \rightarrow (NH_4)_2SO_4 + CaCO_3 \quad ...(2.56)$$

In this process, gypsum comes from two different sources namely, (a) Natural gypsum which comes from gypsum ore, and (b) By-product gypsum which comes from wet process of phosphoric acid. Gypsum may be in the form of (i) calcium sulphate monohydrate ($CaSO_4.H_2O$), (ii) calcium sulphate dihydrate ($CaSO_4.2H_2O$), and (iii) calcium sulphate hemihydrate ($CaSO_4.½H_2O$). However, calcium sulphate dihydrate ($CaSO_4.2H_2O$) is the preferred form of gypsum used for manufacturing of ammonium sulphate.

While $(NH_4)_2SO_4$ is water soluble, the solubility of $CaSO_4.2H_2O$ is having very low solubility (only 0.25%, sparingly soluble), however, it works satisfactorily to produce ammonium sulphate. The principal mechanism involved in this process is explained below:

$$CaSO_4 \rightleftharpoons Ca^{2+} + SO_4^{2-} \quad ...(2.57)$$

$$(NH_4)_2CO_3 \rightleftharpoons 2\,NH_4^+ + CO_3^{2-} \quad ...(2.58)$$

In solution, four types of ions namely, Ca^{2+}, SO_4^{2-}, NH_4^+ and CO_3^{2-} are present. To get forward reaction, the Ca^{2+} concentration should be lowered for dissolution of $CaSO_4$. Based on the solubility product, Ca^{2+} ion concentration is reduced due to precipitation of $CaCO_3$ and removed from the system. Precipitation takes place when ionic product exceeds the solubility product and $CaCO_3$ is precipitated. As more and more $CaCO_3$ is removed, more $CaSO_4$ is dissolved and more $(NH_4)_2SO_4$ is produced. The principles used here are (i) Chemical equilibrium, and (ii) Solubility product principle. After filtering the insoluble $CaCO_3$ from the system $(NH_4)_2SO_4$ is produced.

2.10.2. Properties of Ammonium Sulphate

(1) Ammonium sulphate [$(NH_4)_2SO_4$] is a white crystalline and free-flowing fertilizer.

(2) It contains only 20.5% N, whole of it is in NH_4^+-form and classified as low analysis fertilizer. Low analysis fertilizers are those fertilizers which contain <25% of major nutrients.

(3) It also contains 23.7% S, present in the form of SO_4^{2-}.

(4) It is the best N-fertilizer as far as physical condition is concerned. It is least hygroscopic material. The CRH value is 81, indicating that it has excellent physical property. Thus, no precaution is needed for storage.

(5) As it is a low analysis fertilizer the cost of storage, transport and application is high compared to urea.

(6) Presently, ammonium sulphate accounts for only a small percentage (5-7%) of the total nitrogenous fertilizers.

(7) It is an acid-forming fertilizer and is suitable for use on neutral to alkaline soils. It is usually not used in acidic soils. The equivalent acidity is 107, indicating that it is more acid producing N-fertilizer than urea.

(8) It is a heat-stable fertilizer and its' decomposition takes place at 536 °C.

2.10.3. Use of Ammonium sulphate

(i) The use of high analysis fertilizers cause deficiency of secondary and micronutrients. Since $(NH_4)_2SO_4$ contains both N (20.5%) and S (23.7%), both in plant available form (NH_4^+ and SO_4^{2-} form), they can supply both N and S to crops.

(ii) It can be widely used on the S-deficient soils. It can be used in oilseed crops and pulses where S requirement is more.

(iii) Presence of S prevents scab-disease in potato.

(iv) It is used in acid loving crops, particularly in tea garden where the soil is mild acidic in nature.

(v) Being free-flowing it can be used in fertilizer mixtures but should not be mixed with lime (as it might make ammonia to volatilize).

(vi) However, the cost of ammonium sulphate increases because of S.

2.10.4. Transformation of Ammonium sulphate in soil

It is an inorganic fertilizer. It can supply N to crops immediately after its application in soil. It also supply S to crops. Upon application, $(NH_4)_2SO_4$ is converted to NH_4^+ and SO_4^{2-} ions as presented in equation 1.59.

$$(NH_4)_2SO_4 \rightarrow 2\ NH_4^+ + SO_4^{2-} \qquad ...(2.59)$$

The transformation of $(NH_4)_2SO_4$ in soil and uptake of nutrients by crops and losses of nutrients is schematically presented in Figure 2.3.

2.11. Ammonium Nitrate (NH_4NO_3) and Calcium Ammonium Nitrate

Ammonium nitrate (NH_4NO_3) contains 34% N, out of which half is in ammonium (NH_4^+) form and another half is in nitrate (NO_3^-) form. It is next to urea as far as total N is concerned. The equivalent acidity of ammonium nitrate is 60. It signifies that it is less acidic than all other nitrogenous fertilizers.

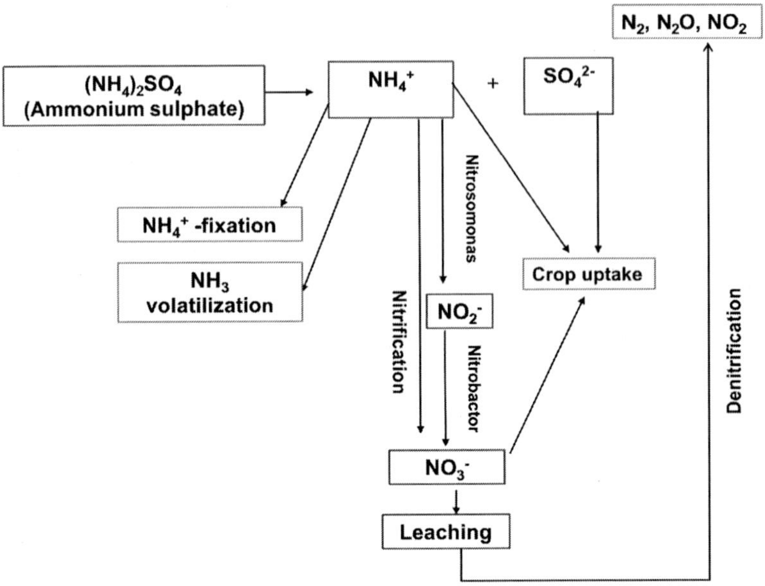

Figure 2.3. Transformation of ammonium sulphate in soil

Ammonium nitrate fertilizer is not used in India because of poor physical properties. However, ammonium nitrate can be mixed with limestone ($CaCO_3$) to produce calcium ammonium nitrate (CAN) (NH_4NO_3+$CaCO_3$,) which is a neutral fertilizer and popularly known as **Kissan Khad** (Farmer's Fertilizer), contains 25% N with equal amounts of ammonium and nitrate-N. The reduction of N in CAN is due to addition of $CaCO_3$ which lowers the hazardous effects of ammonium nitrate. The acidifying property of NH_4^+ is buffered by the $CaCO_3$ in CAN and is therefore very useful as a source of N. In India, CAN is produced/ manufactured by Hindustan Fertilizer Limited, Rourkella and National Fertilizer Limited (NFL), Nangal, Punjab.

2.11.1. Problems of NH_4NO_3

(a) Hygroscopicity

The physical property of ammonium nitrate is very poor as it is a very hygroscopic fertilizer having CRH value of 59 which is very low compared to urea (72) and $(NH_4)_2SO_4$ (81) and NH_4Cl (76). Thus, it can absorb more moisture from atmosphere.

An ideal fertilizer should possess the followings characteristics:

(i) It should be non-hygroscopic and CRH value should be as high as possible.

(ii) It should be free-flowing, and should not form cakes or lumps or aggregates.

(iii) Cost of storage, transport and application should be low. However, NH_4NO_3 is most hygroscopic in nature, thus for storage it needs moisture-proof bags which increase extra cost.

(iv) Moisture in excess leads to formation of lumps or cakes or aggregates. Under such conditions, it is necessary to break the lumps/cakes/aggregates to uniform size or granule for transportation and uniform application in the field. It has to borne in mind that for soluble fertilizers, about 1 mm granule size particle is used for uniform application.

(b) Solubility

Ammonium nitrate is highly water soluble and subjected to losses. It is less resistant to leaching as it contains half of its N in NH_4^+ form, which can remain adsorbed on the exchange complex. When ammonium nitrate or CAN is applied in lowland condition, loss of NO_3^- due to leaching is more because negatively charged NO_3^- ion cannot be retained by negatively charged clay particles. Under lowland situation, the NO_3^- will go to reduced layer, where it will be converted to gaseous N (N_2, NO_2, N_2O, *etc.*). Thus, ammonium nitrate is not used in wetland condition. Ammonium nitrate can be used as solution form because of high solubility along with irrigation water.

(c) Decomposition

Ammonium nitrate is susceptible to decomposition at high temperature around 45 °C and causes explosion as shown in equation 2.60.

$$NH_4NO_3 \xrightarrow[\text{Decomposition}]{45\ ^\circ C} N_2O + 2\ H_2O \qquad ...(2.60)$$

The nitrous oxide (N_2O) gas favours explosion. It increases the intensity of fire. The N_2O is also toxic. It is called laughing gas. It causes hysteric symptom. In normal condition, there is no problem of explosion. It only occurs when the condition is unfavourable *i.e.* if the temperature is high (> 45 °C) and pressure is high. It is also explosive when easily oxidizable materials are present near the storage. Thus, presence of carbonaceous materials such as straw, cotton, *etc.* will accelerate the fire. Therefore, it must be stored at (i) low temperature, (ii) away from electric wires (chance of electric short circuit be there at high temperature), and (iii) away from carbonaceous materials.

(d) Change of crystal geometry with temperature

At ordinary temperature, NH_4NO_3 is cubic shape (a = b = c). It undergoes geometrical changes under variable temperature and pressure. Thus, metamorphosis of crystal shape takes place from cubic to rhombic and then to ortho-rhombic. From ortho-rhombic it again changes its shape to cubic.

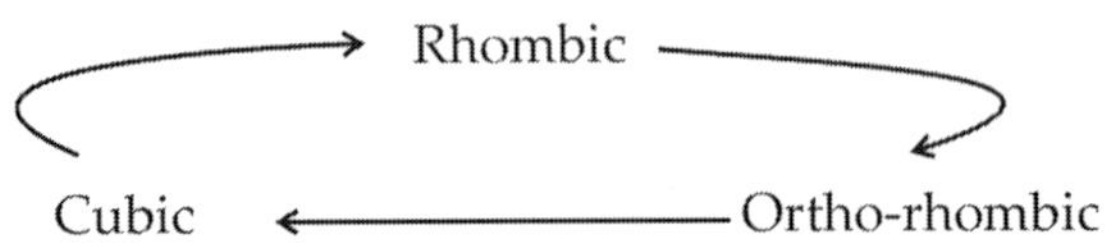

When crystal geometry is changed, volume is also changed. Because of this change, it creates problem during storage. To prevent this problem it should be mixed with $CaCO_3$. The $CaCO_3$ lowers the hazardous effect of NH_4NO_3. However, under prolong storage mixture of NH_4NO_3+$CaCO_3$ (CAN) also undergoes some changes because of the possibility of the reaction of $CaCO_3$ with NH_4NO_3 when it absorbs water from the atmosphere as in the equation 2.61.

$$2\,NH_4NO_3 + CaCO_3 \rightarrow (NH_4)_2CO_3 + Ca(NO_3)_2 \qquad \text{...(2.61)}$$

The above reaction is called double decomposition. Here, two hygroscopic products are formed. Thus, hygroscopicity is further increased. But, when CAN is applied immediately after bagging/marketing, then there will be no such problem.

When $CaCO_3$ is mixed with NH_4NO_3, there is no loss of NH_3 volatilization due to mixing. Since $CaCO_3$ is less alkaline compared to lime (CaO) and contains CO_2 which is acidic in nature and reduces the alkalinity, which in turn, does not allow NH_4NO_3 to loss as NH_3 volatilization.

2.11.2. Manufacturing of Process of NH_4NO_3

Ammonium nitrate can be produced by following two methods.

(a) Ammonium nitrate is manufactured by neutralizing nitric acid with ammonia.

$$HNO_3 + NH_3 \rightarrow NH_4NO_3 \qquad \text{...(2.62)}$$

Nitric acid is an important intermediate compound in manufacturing of nitrogenous fertilizer. It is produced by **Ostwald process** where NH_3 is first oxidized to form nitric oxide (NO), followed by oxidation of NO to NO_2 (nitrogen dioxide) and conversion of NO_2 to nitric acid (HNO_3). This HNO_3 is then reacted with NH_3 to form NH_4NO_3 as in equation 2.62.

$$4\ NH_3 + 5\ O_2 \rightarrow 4\ NO + 6\ H_2O \qquad ...(2.63)$$

$$2\ NO + O_2 \rightarrow 2\ NO_2 \qquad ...(2.64)$$

$$4\ NO_2 + O_2 + 2\ H_2O \rightarrow 4\ HNO_3 \qquad ...(2.65)$$

(b) Nitric acid can also be prepared by Birkeland and Eyde process (Electric arc method)

$$\underset{\text{(Air)}}{N_2 + O_2} \xrightarrow[\text{(3000-3500 °C)}]{\text{Lightning or Electric arc}} 2\ NO \qquad ...(2.66)$$

$$2\ NO + O_2 \rightarrow 2\ NO_2 \qquad ...(2.67)$$

$$3\ NO_2 + H_2O \rightarrow 2\ HNO_3 + NO \qquad ...(2.68)$$

2.12. Ammonium Chloride (NH_4Cl)

2.12.1. Properties of NH_4Cl

(1) Ammonium chloride (NH_4Cl) is the least preferred N-fertilizer. Like $(NH_4)_2SO_4$, it is also called as low analysis fertilizer. Thus, it is not popular as N-fertilizer by farmers as in case of ammonium sulpahte and CAN.

(2) It contains 25% N, whole of which is in NH_4^+-form.

(3) The CRH value of NH_4Cl is 76. It is moderately hygroscopic.

(4) The equivalent acidity is 127, indicating that it causes more acidity than that of urea, ammonium sulphates and ammonium nitrate. The ammonium part of the N-fertilizers is most important for determining or producing soil acidity.

(5) Ammonium chloride sublimes, it signifies that ammonium chloride can be converted or transformed from solid stage to vapour stage without passing through liquid state or no melting or boiling point.

(6) The anion part of the N-fertilizer plays an important role in producing soil acidity which follows the order:

$NH_4Cl > (NH_4)_2SO_4 > NH_4NO_3$ or $Cl^- > SO_4^{2-} > NO_3^-$

The chloride component of ammonium chloride reacts with calcium in soil to produce highly soluble calcium chloride ($CaCl_2$) which is liable to be lost through leaching. For long-term use of NH_4Cl, one has to apply lime (CaO) to neutralize acidity.

(7) Chloride (Cl^-) is an anion which is most toxic. Therefore, the crops which are very susceptible (*e.g.* tobacco, tea, horticultural crops, *etc.*) to Cl^- toxicity,

NH_4Cl is not used. For other crops like wheat, rice, millet it gives equal response to ammonium sulphate. In saline soil, it must not be used. It increases salt index of soil and osmotic pressure. In alkaline soil, it can be used because it has high equivalent acidity and Cl^- concentration in alkaline soil is less. Thus, by application of NH_4Cl in alkaline soil, the pH can be reduced.

(8) It is highly water soluble, thus susceptible to leaching and run-off losses.

2.12.2. Manufacturing Process of NH_4Cl

Ammonium chloride is produced by two processes:

(i) ***Direct neutralization process***: Ammonia is neutralized by reacting with hydrochloric acid to produce ammonium chloride.

$$NH_3 + HCl \rightarrow NH_4Cl \qquad ...(2.69)$$

Production of NH_4Cl by this method is not common as it is not cost effective.

(ii) ***Modified Solvays process***: Method used by the manufacturer, known as the modified Solvays process. The original Solvays process is used for manufacturing of sodium carbonate (soda ash). The modified Solvays process is basically a two-step process. In the first step, ammonia (NH_3) reacts with carbon dioxide (CO_2) to produce ammonium bicarbonate (NH_4HCO_3). In the second step, ammonium bicarbonate reacts with sodium chloride solution (15% NaCl cold brine solution) to produce sodium bicarbonate ($NaHCO_3$) and ammonium chloride (NH_4Cl) (two salts, hence the name dual salt process).

a) Reaction between NH_3 and CO_2 to form NH_4HCO_3

$$NH_3 + CO_2 + H_2O \rightarrow NH_4HCO_3 \qquad ...(2.70)$$

b) Reaction between NH_4HCO_3 and NaCl to form insoluble $NaHCO_3$

$$NH_4HCO_3 + NaCl \rightarrow NH_4Cl + NaHCO_3 \qquad ...(2.71)$$

Sodium bicarbonate is insoluble at low temperature but soluble at high temperature. Thus, by cooling the mixture at temperature of 10 °C (5-15 °C) sodium bicarbonate is precipitated and then separated by filtration. Ammonium chloride solution is then heated to saturation → cooled → crystallized → dried → coated → bagged → marketed.

It may be noted that $KHCO_3$ cannot be prepared by this method, but $NaHCO_3$ can be produced by this method. The reason behind this is that while $NaHCO_3$ is insoluble at low temperature, $KHCO_3$ is highly soluble and thus not produced by this method.

While, in modified Solvays process, the reaction is stopped at $NaHCO_3$ stage and filtered off, the original Solvays process, NaHCO is filtered off and then ignited within the reactor to form Na_2CO_3 as in reaction given below:

$$2\,NaHCO \xrightarrow{\text{ignition}} Na_2CO_3 + CO_2 + H_2O \qquad ...(2.72)$$

The fundamental differences of preparation of urea (NH_2CONH_2), ammonium sulphates [$(NH_4)_2SO_4$] and ammonium chloride (NH_4Cl) are given in Table 2.2.

Table 2. Fundamental differences between urea (NH_2CONH_2), ammonium sulphates $(NH_4)_2SO_4$ and ammonium chloride (NH_4Cl) preparation

	Urea (NH_2CONH_2)	**Ammonium sulphate [$(NH_4)_2SO_4$]**	**Ammonium chloride (NH_4Cl)**
1.	Excess of NH_3 is used, stoichiometric ratio of NH_3:CO_2 is 2:1 (but in practice NH_3:CO_2 is 4:1).	Excess of NH_3 is used, stoichiometric ratio of NH_3: CO_2 is 2:1.	Controlled NH_3 is used, stoichiometric ratio of NH_3: CO_2 is 1:1.
2.	Intermediate product is ammonium carbamate (NH_4COONH_2).	Intermediate product is ammonium carbonate [$(NH_4)_2CO_3$].	Intermediate product is ammonium bicarbonate (NH_4HCO_3).
3.	Reaction involved are: (a) $2\,NH_3+CO_2 \rightarrow NH_4COONH_2$ (b) $NH_4COONH_2 \rightarrow NH_2CONH_2+H_2O$	Reaction involved are: (a) $2\,NH_3+CO_2+H_2O \rightarrow (NH_4)_2CO_3$ (b) $(NH_4)_2CO_3+CaSO_4 \rightarrow (NH_4)_2SO_4+CaCO_3$	Reaction involved are: (a) $NH_3+CO_2+H_2O \rightarrow NH_4HCO_3$ (b) $NH_4HCO_3+NaCl \rightarrow NH_4Cl+NaHCO_3$

Chapter 3

Phosphatic Fertilizers

Phosphorus (P) is the second most important essential nutrient limiting crop/ plant growth, next only to nitrogen (N). Phosphorus is an important component of nucleic acid like deoxyribonucleic acid (DNA) and ribonucleic acid (RNA). It is needed for energy storage and transfer, which is the single most important essential function of P. Phosphate compounds act as "energy currency" within plants. The most important common P energy currency is adenosine diphosphate (ADP) and adenosine triphosphate (ATP). The ATP transfers energy for synthesis of ADP. Life is not possible without P. It is the source of energy that powers practically all energy-requiring biological processes in plants. Phosphorus is present in plant as phospholipid such as lecithin and cephalin which appears to be involved in the structural framework of protoplasm. Phosphorus is involved in many important processes such as photosynthesis, cell division, tissue development, and growth in plant. Requirement of P by the crops is less, compared to N and potassium (K). It is present in plant tissues. It is reported that for 5-6 t ha^{-1} of wheat grain yield, it removes about 100-120 kg N, 20-30 kg P_2O_5 and 150-180 kg K_2O from soil.

Phosphorus is 11th most abundant element on the earth crust. However, it is the least mobile element among the major plant nutrients. Its complex chemistry in soil makes it highly deficient nutrient element in most of the arable soils across the globe. Only 0.1% of the total P in the soil is available for uptake by plants because of its fixation as various insoluble compounds depending upon the soil pH. About 43% of the world's arable soil is P deficient which limits crop growth and production (Roy *et al.*, 2018). In the Indian context, about 80% of the soils are reported to have low to medium available P content (Muralidharudu *et al.*, 2011). Thus, application of P-fertilizers to maintain the soil fertility and sustain crop productivity becomes imperative in today's input intensive agriculture.

3.1. Historical Development P-fertilizers

Henning Brandt (in **1669):** The element phosphorus (P) was synthesized by the German merchant and Alchemist **Henning Brandt** who called it *"The Cold Fire"*. The term "phosphorus" comes from *"Phosfluorences"*. It has two words. *Greek "Phos" = light; and Phero = I bear*. Thus, P is an element that gives or bears some light. That time it was not known that one day P will be treated as the most important macronutrient for plant.

Theodore de Saussure (in 1804) a French, reported the importance of P in agriculture. He wrote a book entitled *"Recherces Chemiques Sur La Vegetation"* means *Chemical Researchers for Plants*. He wrote that phosphorus is an important element for plant growth in this book.

C. Sprengel (in 1839) discovered the essentiality of P in crops/plants.

Justus von Liebig (in 1840), a Physician and Professor of Chemistry at University of Giessen, Germany, reported that plants take phosphorus in inorganic ionic form or mineral form. He gave the ***Mineral Nutrition Theory of Plants.*** He suggested that instead of applying bone meal or other P containing materials as such, the materials should be first reacted with inorganic acids and then applied. He recommended to replenish soil with a product prepared by treating powdered bones with sulphuric acid and found them to be more effective than untreated bones when sprayed this solution after dilution to the fields. This is because it increases the availability of P present in bone meal. Mineral acid decomposes bone meal and releases mineral phosphate which is then taken by plants. Earlier guano, fish and bone were used widely in increasing crop growth. At that time, it was not known that these materials supply P that means without knowing about P, these materials were used. This theory was published in ***BAAS (British Association for the Advancement of Science)*** in 1840. Liebig was invited by BAAS and he delivered a lecture on it and finally they accepted its benefits to plants. This is known as ***"Mineral Nutrition Theory of Plants"***.

The development of the modern phosphate fertilizer industry began with the demonstration by **Justus von Liebig** who reported that the fertilizing value of bone meal could be made more available for plant nutrition by treating it with H_2SO_4. **Justus von Liebig** is recognized as the "**Father of Agricultural Chemistry". Liebig** is also recognized as the "**Founder of Chemical Fertilizer Industry"** for his contribution of using bone meal for production of superphosphate by reacting with H_2SO_4.

Liebig also gave the "**Law of Minimum**" in 1862 which gives a simple but logical guide for predicting crop response to fertilization. The ***"Law of the Minimum" states that if one of the essential plant nutrients is deficient, plant growth will be poor even when all other essential nutrients are abundant, while it will give response by application of that deficient nutrient***. Liebig **Law of Minimum** dominated the thinking of agricultural workers for a long time thereafter and it has been of universal importance in soil fertility management.

John Bennet Lawes (in 1842) produced superphosphate. Prefix "super" is used for "superior". This is the first fertilizer produce in a factory. This gentleman used **Coprolites (Fossilized dung**) for the production of superphosphate. This material (coprolites) contains P and it is known as **phosphorus dung**. He used this material as superphosphate at the **Rothamsted Agricultural Experimental Station** (RAES) in England. Shortly after this, in 1842, **J.B. Lawes** patented a process by which rock phosphate was acidulated with H_2SO_4. He began the commercial production of this material in England in 1843. This was the beginnings of the multimillion dollar fertilizer phosphate industry which uses essentially the same principle for the manufacturing of soluble phosphates employed by J.B. Lawes over a long period of time (since 1842).

3.2. Chronological Development of Phosphatic Fertilizers in India

Chronological production of phosphatic fertilizers as developed by different manufacturers in India is presented in Table 3.1. In India, the first fertilizer was produced in 1906, which happens to be single superphosphate (SSP). Incidentally, the SSP is the first fertilizer produced in most of the countries. The SSP is the first fertilizer material produced across the world, because of the simplicity of its production. It is produced by treating the P-containing materials with inorganic or mineral acids and then concentrating and grinding them into SSP product. In India, EID (Parry) Ltd. for the first time produced SSP at Ranipet about 120 km away from Chennai. This was produced by treating fresh bone meal with H_2SO_4.

In India, ammonium phosphate (AP) was first produced in 1960 by Fertilizers and Chemicals Travancore Ltd. (FACT), Udyogamandal (Alwaye), Kerala. Nitrophosphate was produced in 1965 by Rashtriya Chemicals and Fertilizer Ltd. (RCFL), Trombay, Mumbai; and diammonium phosphate (DAP) was produced for the first time in 1967 by Gujarat State Fertilizers and Chemicals Ltd. (GSFC), Baroda. In 1968, three high-analysis phosphate fertilizers namely, triple superphosphate (TSP), urea ammonium phosphates (UAP) and NPK complex fertilizers arrived in the market. The TSP was produced by Dharamsi Morarji Chemical Co. Ltd., Ambernath; while UAP was produced by Coromandel Fertilizers Ltd., Vizag (Visakhapatnam) and NPK complex fertilizers by RCFL, Trombay, Mumbai.

Table 3.1. Chronological production of phosphatic fertilizers in India

Phosphatic fertilizers	Year	Name of first manufacturer
Single superphosphate (SSP)	1906	EID-Parry (India) Ltd., Ranipet
Ammonium phosphate	1960	The Fertilizers and Chemicals Travancore Limited (FACT), Udyogamandal, Kerala
Ammonium phosphate sulphate (APS)	1960	The Fertilizers and Chemicals Travancore Limited (FACT), Udyogamandal, Kerala
Nitrophosphate	1965	Rashtriya Chemicals and Fertilizers Limited (RCFL), Trombay, Mumbai
Diammonium phosphate (DAP)	1967	Gujarat State Fertilizer and Chemicals Limited (GSFC), Baroda
Triple superphosphate (TSP)	1968	Dharamsi Morarji Chemical Co. Ltd., Ambernath
Urea ammonium phosphate	1968	Coromandel Fertilizers Ltd., Vizag
NPK complex fertilizers	1968	Rashtriya Chemicals and Fertilizers Limited (RCFL), Trombay, Mumbai

Source: Narayanasamy, G., Arora, B.R., Biswas, D.R. and Khanna, S.S (2009) Fertilizers, Manures and Biofertilizers. In: *Fundamentals of Soil Science*. N.N. Goswami *et al*. (Eds.), Indian Society of Soil Science, New Delhi, pp. 579–622.

Potassium nitrate (KNO_3) is also known as **Salt peter** or **Chilean salt**. Although KNO_3 is produced in Chile naturally before production of superphosphate (SSP), but it is a natural product, not an industrial product. It occurs in nature as a mineral commonly called nitre. It is a source of N, from which it derives its name. Potassium nitrate is one of several N-containing compounds collectively referred to as saltpeter. As far back as 1845, Chilean saltpeter deposits were exploited in Chile and California. Major uses of potassium nitrate are in fertilizers, rocket propellants and fireworks. It is one of the major constituents of gunpowder (black powder).

3.3. Raw Materials for Manufacturing of Phosphatic Fertilizers

There are two important raw materials required for production of large number of phosphatic fertilizers. These are sulphur (S) and rock phosphate (RP) or phosphate rock (PR).

3.3.1. Raw Sulphur

Sulphur is an essential raw material for many chemical industries and is essentially used for the production of sulphuric acid (H_2SO_4) which in turn is used for the production of fertilizers, textiles, dyestuffs, pickling and galvanising of steel, storage batteries, refining of petroleum, explosives and other acids. Sulphur is present in the earth surface in various forms like elemental sulphur (S), sulphate form (SO_4^{2-}), sulphide minerals (FeS_2, ZnS, CuS, *etc.*), and sulphur present in different complexes. Out of these, the most important one is elemental S because this is the easiest material to handle. This is used to produce H_2SO_4. If elemental S is not available, then metal sulphide such as pyrites (FeS_2) or ZnS, CuS, *etc.* can be used.

The distribution of S is not well distributed across the world. It is concentrated in five countries namely, China, USA, Russia, Canada and Saudi Arabia. During 2017, the total world production of sulphur (Table 3.2) was 85.2 million tonnes (Mt), out of which China, USA, Russia, Canada and Saudi Arabia produced 20.0, 9.3, 7.0, 6.8, 3.9 Mt of sulphur, respectively (FAI, 2019). China is the leading producer of sulphur in all forms and leading producer of pyrites. Nearly one-fourth of the total production comes from China.

Table 3.2. Production of sulphur by the major countries in 2017

Country	Production (Million tonnes, Mt)	Share to world production (%)
China	20.2	23.7
USA	9.3	10.9
Russia	7.0	8.3
Canada	6.8	8.0
Saudi Arabia	3.9	4.6
Others	38.0	44.5
World total	85.2	100

Source: FAI (2019) Fertiliser Statistics 2018-19. 64th Edition, The Fertiliser Association of India, New Delhi.

In its native form, sulphur is a yellow crystalline solid. In India, though there is no mineable elemental S reserves available at present, but some native S-deposit has been reported in Puga Valley of Leh, Ladak (Indian Minerals Yearbook, 2019c). The grade of the deposit ranges from 9 to 24% of S. Small occurrences of native S-deposits are also reported from Barren Island of Bay of Bengal. Sulphur along with hot springs is reported from various parts of Chamoli, Rudraprayag, Uttarkashi districts in Garhwal and Kumaun divisions of Uttarakhand. In Andhra Pradesh, native S-reserve occurs in granular form with clay and silt in coastal areas of Krishna and East Godavari districts.

Pyrites is naturally occurring mineral comprised of the elements iron and sulphur (FeS_2). Pyrites is a fairly ubiquitous mineral and it occurs most commonly in sedimentary rocks. Pyrites has a brass yellow colour, brownish black streak, metallic lusture and occurs as cubic crystals. Pyrites was used as a substitute for elemental S in the manufacture of H_2SO_4. In India, pyrites was being mined by Pyrites, Phosphates and Chemicals Ltd (PPCL) who had two production units located at Amjhore (Bihar) and Saladipura (Rajasthan). However, there is no production of pyrites since 2003 because of closure and hiving off of these two units by the Government.

Sulphide occurs naturally in mineral ores, oil and coal deposits. Natural waters containing elevated concentrations of hydrogen sulphide (H_2S) are used for therapeutic baths and have been consumed for medical purposes. Hydrogen sulphide, which exists as a colourless gas under normal conditions, has a characteristic odour of rotten eggs and occurs naturally in coal, natural gas, fuel oil, volcanic gases and S springs and lakes. Hydrogen sulphide is also a central component in the S cycle, the biogeochemical cycle of S on earth. Sulphides form is an indispensable link in the S cycle in nature. Petroleum refineries and gas processing plants extract H_2S when making "clean fuels" and use it as a feedstock to produce S. In India, the domestic production of elemental S is limited to by product recoveries from petroleum refineries and fuel oil used as feedstock for manufacturing fertilizer. Tar sands (oil sands) formations contain about 10% bitumen with high H_2S content. The sulphide ores contain S and during the production of metal from sulphide ores, S is released as SO_2 which is used to produce H_2SO_4 which contains about 32.7% of S and contributes in the industries.

3.3.2. Mining of Sulphur

Elemental sulphur is obtained from ores by conventional mining or as a by-product of natural gas processing, crude refining, tar sand processing and stack gas clean-up (recovered sulphur). Recovered S production accounted for over 98% of world elemental S production.

For mining, usually three concentric pipes are used. The outermost pipe contains superheated water, which melts the S, and the innermost pipe is filled with hot compressed air, which serves to create foam and pressure. The resulting S foam is then expelled through the middle pipe. This process produces S with 99.5% purity, and it needs no further purification. This method of S production on a commercial scale is operated in Brazil and Mexico. Elemental or native S is mined in China, Poland and Russia.

3.3.3. Sulphuric Acid Production

Currently, the global production of sulphuric acid (H_2SO_4) stands at more than 270 Mt (FAI, 2019). About 60% of the world's production of H_2SO_4 is based on elemental S, 25% is pyrites and the rest on other sources. So far, there are five countries that rank as the leading producers of H_2SO_4. Leading the chart is China, whose production output stands at more than 74 Mt. The USA sits second with more than 37 Mt, India produces 16 Mt, Russia accounts for 14 Mt, and Morocco produces 7 Mt of the world output. Generally, these account for approximately 61.5% of the global production of H_2SO_4. Currently, the fertilizer industry consumes more than 55% of the world output. Further, nearly 50% of the total H_2SO_4 produced world over is used for the manufacture of phosphatic fertilizers of which 75% are in the form of SSP. It is anticipated that the global demand of H_2SO_4 is rising, while there is a need for an increase in production to meet this demand.

India is one of the major producers of H_2SO_4 in the world. So far, India utilizes the majority of its H_2SO_4 in the production of fertilizers. The process involves the acidulation of rock phosphates, as well as the production of ammonium sulphate [$(NH_4)_2SO_4$]. Other industrial applications of H_2SO_4 include petroleum refining, steelworks and other inorganic chemicals. Currently, there are more than 65 sulphuric acid plants spread across India. The major raw materials used in the manufacture of H_2SO_4 include elemental sulphur, hydrogen sulphide, pyrites, *etc.* Nearly all factories in India rely heavily on elemental sulphur as the major source of raw material.

3.3.4. Roasting of Elemental Sulphur for Production of Sulphuric Acid

There have been evolutions in the methods for the production of H_2SO_4. However, there are two major processes for its commercial production namely,

(i) lead chamber process, and (ii) contact process. Irrespective of the difference in the processes, the manufacture of H_2SO_4 is generally based on the following principles:

The extraction of pure elemental sulphur from its ore.

- The conversion of elemental sulphur (S) by roasting into sulphur dioxide (SO_2) in the presence of oxygen.
- The conversion of sulphur dioxide (SO_2) into sulphur trioxide (SO_3).
- The conversion of sulphur trioxide (SO_3) into sulphuric acid (H_2SO_4).

In practice, elemental S is burnt (roasted) in presence of dry air at a temperature of 950-1100 °C to produce sulphur dioxide (SO_2) which is converted to sulphur trioxide (SO_3) through catalytic oxidation. The SO_3 is subsequently absorbed in water to form dilute H_2SO_4. The diluted H_2SO_4 product is then concentrated to usually 98% concentration.

3.4. Sources of Phosphate

The original source of P in the early manufacturer of phosphatic fertilizer was bones or bone meals, but the supply of bones was soon exhausted. Today, the only important source of fertilizer and industrial P is phosphate rock (PR) or rock phosphate (RP) ores mined in various parts of the world.

3.4.1. Apatite

The term apatite is a group of phosphate minerals and named by **German Geologist Abraham Werner** in 1786. Rock phosphate occurs in nature as mineral phosphate in deposits which is known as apatite (phosphate-bearing mineral) along with other accessory minerals such as quartz, silicates, carbonates, sulphates, sesquioxides, *etc.* It is the most abundant crystalline phosphate mineral found as an accessory mineral in practically all kinds of igneous rocks. It also occurs in metamorphic rocks and as a secondary mineral in phosphatic rocks of sedimentary origin. It is a group of phosphate minerals, usually referring to fluorapatite [$Ca_{10}(PO_4)_6F_2$], hydroxyapatite [$Ca_{10}(PO_4)_6(F,OH)_2$] and chlorapatite [$Ca_{10}(PO_4)_6(F, Cl)_2$]. Among these, fluorapatite is the most common variety of apatite and also a secondary source of fluorine.

3.4.2. Rock Phosphate or Phosphate Rock

Rock phosphates are sedimentary phosphatic deposits comprising fine grained mixture of various calcium phosphates, most important being hydroxyapatite, carbonate apatite, fluorapatite and their solid solutions. About 80% phosphate production in the world is derived from rock phosphates (phosphorite) containing one or more phosphatic minerals, usually calcium phosphate of sufficient purity and quantity to permit its use directly or after concentration in manufacturing commercial products.

The term rock phosphate (RP) or phosphate rock (PR) may be defined as any rock containing large proportion of phosphate minerals that can be used as such for commercial manufacture of fertilizers, direct application to soil or production of industrial phosphorus compounds. Usually it is produced by mining and beneficiation. About 150 minerals containing more than 1% P_2O_5 occur in nature. But minerals belonging to two groups, namely '**Apatites**' and "**Aluminous phosphates**" alone are of commercial importance. In most of commercial grade rock phosphate, the P_2O_5 content is around 32% (w/w).

Rock phosphate is one of the basic raw materials needed for manufacturing of all phosphatic fertilizers like single superphosphate (SSP), triple superphosphate (TSP), nitrophosphates and intermediate products such as phosphoric acid (H_3PO_4) and elemental phosphorus (P).

India is deficient in apatite and rock phosphate availability. In case of apatite, the country is fully dependent upon imports, while the rock phosphate production is mainly from two states namely, Rajasthan and Madhya Pradesh and to some extent from Purulia, West Bengal.

Sometimes, rock phosphate is known as **phosphorite**. A phosphorite is a phosphate deposit of economic interest, directly or indirectly of sedimentary origin.

The phosphate is usually in the form of small **pellets** or **pebbles** cemented by $CaCO_3$. In some cases, the cementing material is leached away, leaving more or less loose pebbles; in others, the whole mass has been compressed to form hard rock.

3.4.3. Phosphate Deposits

Phosphate deposits may be classified as (i) Residual, (ii) Replacement, and (iii) Sedimentary.

(i) **Residual phosphates** such as **Tennessee brown phosphates** are derived from beds of phosphatic limestone.

(ii) **Replacement phosphates** such as **Nauru phosphates** are limestone that have been phosphatized or changed to calcium phosphate by the action of phosphate-bearing solutions derived from guano or other organic sources. Nauru's phosphate deposit is the result of thousands of years of bird droppings, also called **guano**. This phosphate-rich deposit lies near the surface, allowing easy strip mining operation.

Nauru is a small island country, in the Pacific Ocean, with an area of only 21 sq. km. It is the smallest state in the South Pacific and third smallest state by area in the world. Until the late 19th century it was a colony by the German Empire. The Europeans soon discovered phosphate deposits and the tiny island became a strip mine, exploited by foreign colonial powers. After it gained its independence in 1968, mining intensified until most of the phosphate had been stripped. In the process of mining the country had rendered its own landscape infertile. Today, the island is a barren wasteland with limestone that covers 80% of the island.

(iii) **Sedimentary phosphate deposits** are those that have accumulated as sediments and later have become interbedded with other rocks, usually limestones and shales. It is generally conceded by geologists that the larger deposits of calcium phosphate, which are the sedimentary deposits, are of organic origin. Such deposits have been laid down in the beds of sea. Sea water carries a trace of phosphates which is drawn upon by marine organisms and stored in their bodies. Upon death, the remains of these organisms accumulate on the floor of the sea, and the phosphatic content is concentrated through the decay of the organic matter and loss of part of its soluble carbonates.

In some cases beds of phosphatic limestone which have been formed in this manner have been raised above the surface of the sea and their phosphate content still further concentrated through leaching of soluble carbonates. It is believed that it was in such a manner that phosphates of the **South-eastern part of the United States** originated. Many bones and teeth of pre-historic fish may still be found in some of these deposits.

3.4.4. Distribution or Occurrence of Rock Phosphates

Rock phosphate is distributed in all over the world. However, like other mineral resources, rock phosphate deposits are also distributed inequitably globally. Though many countries produce rock phosphates, it has been found that the distribution of rock phosphate reserves is concentrated only few countries mainly situated at the Mediterranean countries. Presently, the world reserves of rock phosphate are about 69 billion tonnes (Indian Minerals Yearbook, 2019a), located mainly in Morocco and Western Sahara (72%), China (5%) and Syria (3%) and Algeria (3%). These countries contribute nearly 83% of the world reserves of rock phosphates. The remaining 17% is located in

Table 3.3. World reserves of rock phosphate by major countries

Country	Reserves of rock phosphate (in Million tonnes, Mt)	
	2017	**2018**
World: Total (rounded)	**70000**	**69000**
Algeria	2200	2200
Australia	1100	1200#
Brazil	1700	1700
China	3200	3200
Egypt	1300	1300
Finland	1000	1000
India*	46	46
Israel	67	62
Jordan	1000	1000
Kazakhstan	260	260
Mexico	30	30
Morocco & Western Sahara	50000	50000
Peru	400	210
Russia	600	600
Saudi Arabia	1400	1400
Senegal	50	50
South Africa	1500	1400
Syria	1800	1800
Togo	30	30
Tunisia	100	100
USA	1000	1000
Uzbekistan	100	100
Vietnam	30	30
Other countries	770	770

#For Australia, Joint Ore Reserves Committee compliant reserves are 81 million tonnes.

**India's total reserves/resources as on 1.5.2015 are 312.67 Mt.*

Source: U.S. Geological Survey, Mineral Commodity Summaries, February 2020; Indian Minerals Yearbook (2019a) *Apatite and Rock Phosphate*. Part-III: Mineral Reviews, 58th Edition, Government of India, Ministry of Mines, Indian Bureau of Mines, Indira Bhavan, Civil Lines, Nagpur.

other countries (Table 3.3). In terms of reserves, Morocco and Western Sahara maintained its position as the World leader with 50 billion tonnes followed by China, with 3.2 billion, USA 1.0 billion, South Africa with 1.4 billion, Jordan with 1.0 billion and Australia 1.2 billion tonnes. These countries contribute nearly 83% of the world assets in rock phosphates.

The world production of rock phosphate decreased slightly by 9% to 232 Mt in 2018 from 255 Mt in 2017 (Table 3.4). China (42%), Morocco (16%), USA (12%), Russia (6%) and Jordan (3%) have been the major producers of rock phosphate.

The recent report by FAI (2019) showed that China, Morocco, USA, Russia and Jordan are the five major rock phosphate producing countries in the world during 2017 contributing 38.8, 15.8, 13.1, 6.4 and 4.2% share to world production (Table 3.5). Amongst other rock phosphates producing countries, important countries are Togo, Tunisia, Senegal, Israel and Algeria. The major producer of phosphoric acid (Table 3.6) by the major countries in 2017 are China, USA, Morocco, Russia and Inda contributing 38.3, 15.6, 12.2, 7.2 and 3.6% of share to world production (FAI, 2019).

3.4.5. Classification of phosphates

Sources of phosphates can be classified into inorganic and organic (Fig. 3.1). While, biomass, bone or bonemeals and guano (dropping of bird excreta) and the major organic phosphates, the major inorganic phosphates are basic slag and rock phosphates.

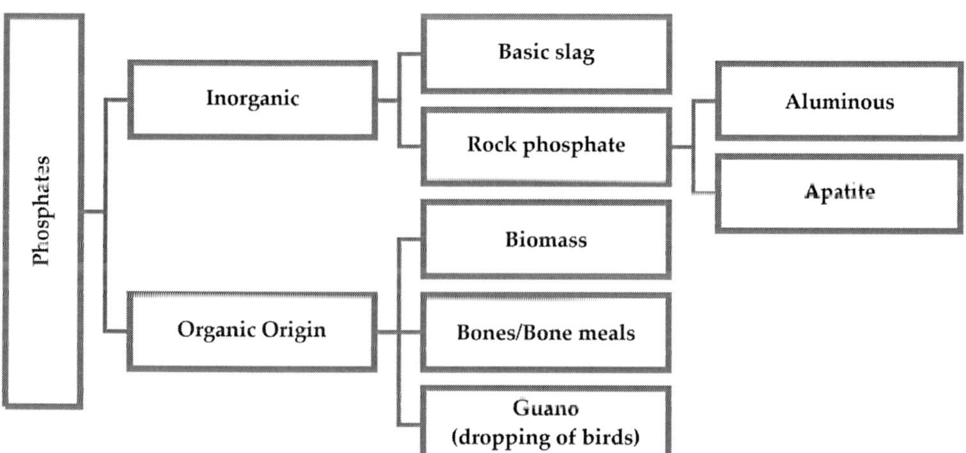

Fig. 3.1. Classification of phosphates

Table 3.4. World production of rock phosphate by major countries

Country	Production of rock phosphate (in Million tonnes, Mt)	
	2017	**2018**
World: Total (rounded off)	255.00	232.00
China	123.13	96.33
Morocco	32.80	37.60
USA	27.90	27.00#
Russia	13.20	13.60
Peru	8.45	10.31
Jordan	8.69	8.02
Saudi Arabia	5.80	6.09
Vietnam	4.59	5.42
Brazil@	5.35	5.10
Israel	2.64	3.03
Tunisia	4.42	2.80
Egypt	2.50#	2.50#
South Africa	2.08	2.06
Kazakhstan	1.83#	1.83#
Senegal	1.39	1.78
India*	1.53	1.25
Algeria	1.11	1.20
Finland	0.98	0.99
Togo	0.73	0.93
Australia	0.94	0.85
Other countries	4.80	2.85

Estimate

@ Including beneficiated and directly shipped material.

* India's production was 1.57, 1.12 and 1.53 Mt, respectively during 2015-16, 2016-17 and 2017-18.

Source: U.S. Geological Survey, Mineral Commodity Summaries, February 2020; Indian Minerals Yearbook (2019a) *Apatite and Rock Phosph*ate. Part-III: Mineral Reviews, 58th Edition, Government of India, Ministry of Mines, Indian Bureau of Mines, Indira Bhavan, Civil Lines, Nagpur.

Table 3.5. Major producer of rock phosphates in the world during 2017

Country	Production (Million tonnes, Mt)	(%)
China	81.0	38.8
USA	27.4	13.1
Morocco	33.0	15.8
Russia	13.3	6.4
Jordan	8.7	4.2
Others	45.4	21.6
World total	208.8	100

Source: FAI (2019)

Table 3.6. Production of phosphoric acid by the major countries during 2017

Country	Production (Million tonnes, Mt)	Share to world production (%)
China	17.8	38.3
USA	7.3	15.6
Morocco	5.7	12.2
Russia	3.4	7.2
India	1.7	3.6
World total	46.4	100

Source: FAI (2019)

The rock phosphate again comprised of (a) aluminous phosphates, and (b) apatites.

3.4.5.1. Aluminous Phosphates

This group of minerals is fairly widespread in nature. The principal aluminum phosphate minerals are:

(i) Crandallite [$CaO.2Al_2O_3.P_2O_5.5H_2O$]

(ii) Wavellite [$4AlPO_4.2Al(OH)_3.9H_2O$]

(iii) Millisite [$2CaO.Na_2O.6Al_2O_3.4P_2O_5.17H_2O$]

Aluminous phosphates as its name suggests, contains aluminium (Al) and phosphate. Beside these, it also contains Na, Mg, Ca and CO_3 due to isomorphous substitution. It may be borne in mind that as in apatites, various substitutions are possible such as Fe for Al is one common substitution. Some of the important places where these materials are found are **Christmas Island** (New Zealand) and **Florida,** USA. It also found in Australia, Nauru and few other places.

Potentiality of Aluminous Phosphate as P Source

- Aluminous rock phosphate is generally used for long duration crops. But the efficiency of these phosphates is very low when applied directly to soil. Thus, to increase the efficiency of aluminous rock phosphate, some physical treatment is given which is known as **Calcination**. Here, rock phosphate is subjected to a thermal treatment under controlled condition, which gives higher availability of P from these sources.
- The contribution of aluminum phosphate to fertilizer usage is very small. Only <5% rock phosphate comes from aluminous form and 95% comes from apatite form.
- Availability of aluminous rock phosphate is also less compared to apatite.
- It is generally agreed that ground rock phosphates is effective only on acid soils (pH <6). This statement applies to apatite rocks, not to calcined aluminium phosphate ores which are effective on neutral or calcareous soils.

3.4.5.2. Apatites

These are basically calcium phosphate and almost the entire quantity of fertilizer P and other industrial P-compounds is produced from these minerals. Commercial rock phosphates generally contain a variant of apatites such as Fluorapatite [$Ca_{10}(PO_4)_6.F_2$] as the primary source of P. Apatite occurs in igneous, metamorphic and sedimentary rock formations. Fluorapatite exists as a part of igneous work. In general, apatite comes from molten magma. The igneous rock undergoes a series of changes and forms metamorphic and sedimentary rocks. Metamorphic rock phosphate is formed in high temperature and pressure. Sedimentary rocks are formed by the accumulation of igneous and metamorphic rocks which undergoes weathering. Due to weathering, rock phosphates are disintegrated and accumulated in the form of sedimentary rocks.

3.5. Properties of Rock Phosphate

The quality of rock phosphates and their availability as source of P to crops depends on the reactivity of rock phosphate, isomorphous substitution and chemical compositions including presence of other accessory minerals.

3.5.1. Reactivity of Rock Phosphate

Apatites occurring in deposits of igneous and metamorphic origin are generally regarded as less reactive, as these materials exist in well developed crystalline form and consequently result in poor performance when applied to soil as source of P. The sedimentary rock deposits, on the other hand, are generally soft materials possessing microcrystalline structure and comprise of younger apatites formed mostly in marine origin. These are of major commercial importance for direct application in the soil. Apatite first comes from magma. As the magma solidified different forms of element/compounds are formed which are known as fluorapatite. This is the first form of apatite which comes first due to solidification of magma.

3.5.2. Isomorphous Substitution

Some chemical changes occur in fluorapatite which is called isomorphous substitution. Due to its isomorphous substitution, phosphate (PO_4^{3-}) is substituted by carbonate (CO_3^{2-}) and ($CO_3^{2-} + F^-$) in the pure fluorapatite crystal structure [$Ca_{10}(PO_4)_6F_2$]. However, full isomorphous substitution has not been encountered anywhere, but partial replacement of PO_4^{3-} by CO_3^{2-} is common. This type of substitution up to varying level leads to the formation of a series of apatites which are called "***Carbonate apatite***" or "***Francolites***". As a result of changes of fluorapatite, it forms to hydroxyapatite, chlorapatite and finally Carbonate apatite (Francolite).

Partial replacement of F^- by OH^- or Cl^- occurs in fluorapatite [$Ca_{10}(PO_4)_6F_2$] structure, resulting formation of hydroxyapatite [$Ca_{10}(PO_4)_6(F, OH)_2$] which is found in phosphates from Nauru Island, Christmass and Brazilian igneous rock and chlorapatite $Ca_{10}(PO_4)_6(F, Cl)_2$ which is found in Sri Lanka. However, there is no evidence that F^- is totally replaced by either OH^- or Cl^-. Therefore, to be more precise one should consider fluor-hydroxyapatite, and fluor-chlorapatite.

The primary substitution takes place in "PO_4^{3-} position". But there are also evidences that substitution in "Ca position" is also possible. But in this case a divalent element replaces a tri-valent element. Thus, there are some changes of geometry or some changes in chemical structure occurs.

McClellan and Lehr (1969) introduced the general empirical formula of the carbonate apatite or fracolite of sedimentary rock phosphate, as obtained from X-ray diffraction (XRD) analysis is given below:

$Ca_{10-a-b}Na_aMg_b(PO_4)_{6-x}(CO_3)_xF_2F_Y$

where, 'a' represents the molar substitution of Na for Ca; 'b' represents the molar substitution of Mg for Ca; 'x' is the molar substitution of CO_3 and F for PO_4. The value of 'x' ranges from 1 to 1.5 in naturally occurring apatites in commercial rock phosphates.

In general, the value of

b = 0 - 4a

a = x - y

y = 0.4a

a, b and y are internally regulated.

when, CO_3 = is zero (x = 0), the compound is known as fluorapatite.

The phosphate content in fluoraptite is more than carbonate apatite. Calcium content also goes down because of the partial substitution of Ca by Na and Mg.

After substitution, the francolite becomes microcrystallined which is bonded by lime itself. Because of microcrystallined structure they have larger surface area (both externally and internally). Thus, the reactivity is more compared to well crystallined materials. Thus, availability of P in francolite increases with the extent of isomorphous substitution.

3.5.3. Chemical Composition of Apatite

All practical purposes, rock phosphate comes from apatite. Since it is a natural occurring mineral, its chemical composition, physical properties are also different from one source to others. The naturally occurring rock phosphates are highly heterogenous in composition. Even within a deposit, the chemical composition is affected by the quantity and quality of accessory minerals present, though the nature of apatite itself may be fairly homogeneous. In fertilizer industry, the material having **more than 30% P_2O_5** is normally used as raw material. Indian rock phosphates are having less than 30% P_2O_5. Therefore, these materials are not used for commercial production of P-fertilizers.

The quality of rock phosphate depends on (i) age of rock, (ii) degree of substitution, and (iii) solubility in organic acids. Depending on these, rock phosphates deposits are classified as high-grade (>30% P_2O_5), medium-grade (25-30% P_2O_5) and low-grade (<25% P_2O_5).

3.5.4. Accessory Minerals

Apatites occur is natural deposits along with some accessory (impurity) minerals. The most common ones are clays, qauartz, silicates, sulphates, sesquioxides, carbonates (mainly calcite and dolomite) and micas. A crude mined ore containing these accessory minerals are usually purified to some extent by a process known as "**beneficiation**" to set **apatite** containing 30-38% P_2O_5.

3.6. Mining of Rock Phosphate

Mining is a process by which rock phosphate containing apatite is obtained from the earth surface. On the surface where rock phosphate is abundant, the extraction is easy. Mining of rock phosphate is very much depended on the geology of apatite. There are two types of mining process depending on the depth of over burden layer. The over burden material may be present below 2 m depth only or sometime many be below 400 m depth. The thickness of over burden material is very important for mining. Depending on this, different methods are followed. If over burden is deep, then underground mining is followed.

(i) **Open cast mining or strip mining:** If the rock phosphate is found within a depth of 30-40 m deep from the surface, then mining is done by open cast mining.

(ii) **Underground mining:** This method is followed, if the depth of over burden layer is below 100-400 m deep, that means when the rock phosphate deposits is present in the lower depth, then the over burden material is first removed. Thereafter, rock phosphate is mined. This process is known as underground mining.

Usually, mining is concentrated to strip or open cast mining. This operation is simpler and cost involved is less. Thus, this method is normally followed. Today most of the rock phosphate comes through open cast mining. In case of underground mining, the cost is much higher because it involves lot of labour and energy for mining. Therefore, usually it is not followed for mining rock phosphate.

Once the rock phosphate material is mined, they are transported to **Sump**. The task of removing overburden materials, digging ore and delivering it to a sump is carried out by huge dragline operations (Fig. 3.2). Normally, a sump is arranged to have two hydraulic guns (2½″ diameter) placed on either side of the sump through dragline. Another smaller gun is used to break up mud balls or other large agglomeration that can not pass through the hydraulic gun. A jet of water is passed through this hydraulic gun to sump and finally a slurry of sump is formed which are then pumped to the washer plant by lifting. Then wet sieving is done generally to split into 3 fractions.

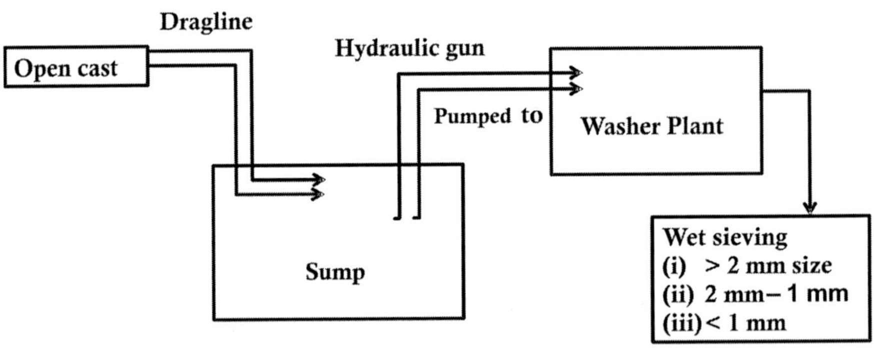

Fig. 3.2. Apatite is dragged through dragline

By wet-sieving, the materials having >2 cm size is first collected and separated. Then 40% material passing through >2 cm sieve but retain on >1 mm sieve (2nd fraction) is collected. Finally the material passing through 1 mm sieve (3rd fraction, <1 mm size) is collected.

Wet-sieving is done for separating different size fraction.

- Fine size (<1 mm) go to the floatation plant
- Over size (>2 mm) go to log washers and hammer mills, and
- Middle size (1-2 mm) are scrubbed and screened by removing clay and sand fractions.

3.7. Beneficiation Technique of Rock Phosphate

A crude mine ore containing accessory minerals like silicates, carbonates, sulphates, quartz, clays and other impurities, is usually purified to some extent by a process known as "beneficiation" to get a apatite-rich rock phosphate containing 30-38% P_2O_5. After beneficiation, the material is then ground into fine powder form so that 90% of the material should pass through 100-mesh sieve for subsequent use as raw material in phosphate industry.

The beneficiation process is a relatively complex procedure that involves:

(i) Wet screening/sieving for separation of different size fraction,

(ii) Hydroseparation, and

(iii) Concentration by flotation.

The process results in several size fractions that include a pebble fraction over 1.19 mm (+16 mesh) in size containing 30 to 35% P_2O_5, and smaller size fractions containing >38% P_2O_5.

The beneficiation is done mainly to remove impurities from the mined ore. Thus, problems of impurities can be reduced to a desired level. In fact any impurities present should be removed. But, definitely the economics come into picture. However, it is very important to see the economics of beneficiation. This indicates that part of the impurities can be allowed in the material up to the limit which does not affect the quality of rock phosphate used as raw material for the industry. There is no rock phosphate which is used directly as such after mining.

Impurities present in rock phosphate can be removed by physically or chemically or by both. Some physical principles or chemical reactions are involved.

3.7.1 Froth Floatation Technique

This method is followed for Florida Pebble, USA. Essentially, it involves some physical techniques or physical principle is used to separate impurities. It utilizes the different specific gravity of the apatites and impurities. A froth is found where certain material is float over the liquid. In this method, certain frothing materials like caustic soda, fatty acids, fuel oil and kerosene are used. The frothing materials have lower density than the liquid. Some materials are attracted by froth and float and some others are not attracted by the froth and sinks. In this technique normally frothing material attract apatite and float, and impurities are deposited on the floor as mud. This technique is possible where the particle size of the apatite is <2 mm size. Particle size >2 mm are unable to float because of higher specific gravity. Here, water is added to powder apatite, mix well. Then frothing material is added, agitated and allowed to settle. The upper portion containing froth plus apatite is then separated from the impurities.

Prior to the discovery of flotation process, the "pebble fraction" was the only phosphate recovered after mining phosphate ore. Together with the slimes, the fine fractions were dumped in mined-out areas, as there was no method known for the separation of small phosphate particles form sand and clay.

The idea of separating phosphate materials from adhering gangue particles by means of flotation was first conceived in 1920. The process employed was patented by **Bradridge and Edser (1925)**. Fine phosphate particles were separated from silica sand by a froth-flotation process. This method uses fatty acid. However, a satisfactory separation of +1 mm +28 mesh is performed by agglomeration of coarse fraction along with fine fractions. After recovering the +1 mm fraction (pebble rock fraction), the remaining material is separated into two fractions by following processes:

(a) **Agglomerate flotation** of the +1 mm, +28 mesh material, and

(b) **Cell flotation** of the –28 mesh + 200 mesh material

The first fraction is separated using a mixture of fatty acid, kerosene, fuel oil and caustic soda. Principally, a mixture containing solids of various specific gravities are suspended in liquid and separated while travelling down a nest of spirals in such a way that the heaviest solids will be concentrated on the inner rim, the lightest will be concentrating towards the outer rim and those of intermediate density remaining on the middle path. Owing to its low specific gravity, rock phosphate tends to concentrate on the outer path.

3.7.2. Double Float Technique

3.7.2.1. Agglomeration Flotation Technique (First step) (<1 mm, 28-mesh size)

In this case, a suspension of materials having 30-35% solid (1 mm size, 28-mesh) is required. A suspension is made using 65-70% liquid.

Some reagents are used like caustic soda, fatty acids, fuel oil or kerosene oil. The whole materials are stirred mechanically so that froth is formed on the surface. Compositions of frothing materials per tonne of ore comprising of caustic soda (NaOH) 0.3-0.6 kg, fatty acid 0.4-1.25 kg, fuel oil 4-5 kg and kerosene. These materials are added proportionality. The whole materials are stirred mechanically so that froth forms on the surface. The froth is capable of keeping the apatite material with them leaving most of the impurities in the container.

The froth materials along with apatites are removed from the container. Then wash with dilute H_2SO_4 to remove oils and traces of fatty acids. After repeated washing with H_2SO_4, the materials containing caustic soda (NaOH) is neutralized. The materials is then dried and stored. The material thus produced is called **fatty acid concentrated** material.

3.7.2.2. Cell Flotation (Second step) (28-200 mesh sieve)

Second stage flotation technique is called cell floatation. Here, fatty acid concentrated material is subjected to other reagents. Certain amines are used as reagents. Then agitation is done and again froth comes out and removed from the container. In this case, impurities come out with the froth and phosphate materials remain in the bottom of the second stage of the container.

3.8. Drying of Rock Phosphate

After removing the impurities a concentrated phosphate (90%) is obtained. This material is then dried at 82 °C. Before drying, moisture content is about 8-15% which is reduced to 2-3% after drying. The drying of phosphate is required to remove water from the solid. Otherwise, it may forms clod or lump which hamper in grinding.

3.9. Grinding of Rock Phosphate

Grinding is done to reduce the particle size. The large particle is ground and a fine material is produced. Generally, the rock phosphate particles are ground so that 90% of the materials pass through 100-mesh sieve. As the number of hole or openings increases, diameter of each aperture reduces.

Note: The term "mesh" is a British count. This count is the mesh number. It means how many openings are there per sq. inch area. A 4-mesh sieve or screen means there are four little square openings across one inch of screen. A 100-mesh sieve or screen has 100 openings per inch, and so on. As the number of the mesh size increases, the size of the openings and thus the size of particles retained by the screen decreases.

Three factors are important for effective grinding.

(a) ***Moisture***: For easy grinding, the moisture content should be 1-3%. There is a steep fall of grinding when moisture content is more than 3%.

(b) ***Temperature:*** It is not a critical factor. However, usually temperature should be maintained at room temperature.

(c) ***Presence of insoluble material***: Presence of insoluble materials like quartz/ silicates affect the grinding process.

3.10. Classification of Phosphate Fertilizers

The original source of P in the early manufacture of phosphatic fertilizer was bones, but the supply of bones was soon exhausted. Today, the only important source of fertilizer and industrial P is rock phosphate. Commercially important fertilizer phosphates are classified as:

1. Thermal-processed phosphates, and
2. Acid-treated phosphates

1. Thermal-processed phosphates are:
   a) Elemental phosphorus, and
   b) Defluorinated phosphates
2. Acid-treated P-fertilizers are by far the more important group. Acid-treated phosphates are classified as:
   a) Phosphoric acid
   b) Superphosphates - Single superphosphate (SSP) and Triple superphosphate (TSP)
   c) Ammonium phosphates - Monoammonium phosphate (MAP) and Diammonium phosphate (DAP)
   d) Ammonium polyphosphates
   e) Nitric phosphates (Nirtophosphates), and
   f) Miscellaneous and new phosphate materials

In all the processes, rock phosphate is the basic material or building block as shown in Fig. 3.3. These does not means each and every rock phosphate is used for using all the products.

3.11. Thermal Processed Phosphates

3.11.1. Elemental Phosphorus Manufacture

If rock phosphate is heated above 1400 °C in the presence of silica and carbon in a reducing atmosphere, elemental P is formed. The reaction is complex, but it can be expressed by the following generalized formula:

$$Ca_3(PO_4)_2 + 3\,SiO_2 + 5\,C \xrightarrow{(1400\ ^\circ C)} 3\,CaSiO_3 + P_2 + 5\,CO\,(\uparrow) \qquad \dots (3.1)$$

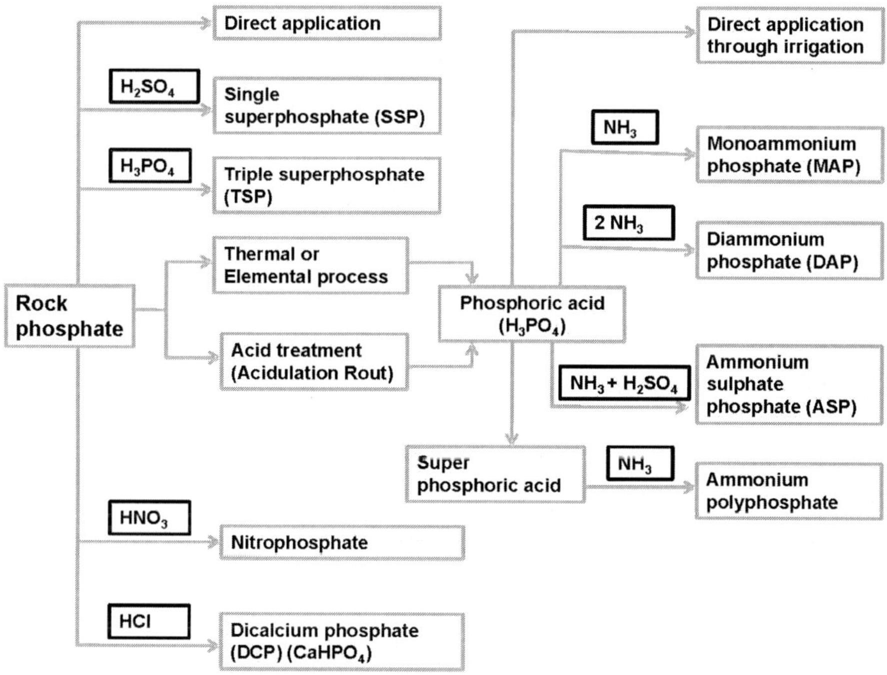

Fig. 3.3. Schematic diagram of different P-fertilizers production using rock phosphate as the key-input (rock phosphate as raw material used for P-fertilizer industry)

The above reaction can also be written as:

$$Ca_{10}(PO_4)_6 + 6\,SiO_2 + 15\,C \xrightarrow[(1400\ ^\circ C)]{heating} 3(3CaO.2SiO_2) + 1.5\,P_4 + CaF_2 + 15\,CO\,(\uparrow) \quad \dots (3.2)$$

Elemental P cannot exist in nature because it is highly reactive. Elemental P comes in contact with O_2 and oxidation takes place and produce P_2O_5.

$$2\,P + 2\tfrac{1}{2}\,O_2 \rightarrow P_2O_5 \quad \dots (3.3)$$

Thus, elemental P is stored in the liquid state until ready for use.

Most of the elemental P is used for preparation of high-purity phosphoric acid (H_3PO_4) for industrial needs, particularly for pharmaceuticals and water conditioners or animal feeds where higher value material is produced.

The anhydride of phosphorus (P_2O_5), also known as phosphorus pentaoxide, is the only oxide of P in fertilizers. The symbol P_2O_5 is commonly called in the fertilizer trade as phosphoric acid or phosphate. Water combines with P_2O_5 to form a series of acids, among which the following three are of chief interest to the fertilizer industry:

(i) $P_2O_5 + H_2O \rightarrow 2\ HPO_3$ (Metaphosphoric acid, MPA) ... (3.4)

(ii) $P_2O_5 + 2\ H_2O \rightarrow H_4P_2O_7$ (Pyrophosphoric acid, PPA) ... (3.5)

(iii) $P_2O_5 + 3\ H_2O \rightarrow 2H_3PO_4$ (Orthophosphoric acid, OPA) ... (3.6)

The first two forms are not stable in nature and third one (H_3PO_4) is stable. First two forms may be stored in the laboratory condition.

Chemists believed that all phosphates can be represented stoichiometrically as combinations of oxides. For example, dicalcium phosphate ($CaHPO_4$) can be represented as $CaO.H_2O.P_2O_5$.

Pyrophosphoric acid ($H_4P_2O_7$) – prefix 'pyro' indicates that it occurs in substances that are subjected to intense heating.

Metaphosphoric acid (HPO_3) – It results when the substance is over heated.

It is known that small amounts of pyro- and meta-phosphates may form in superphosphate when it is subjected to elevated temperature during the period of drying.

3.11.2. Defluorinated Phosphates

When rock phosphate is heated in a gas or oil-fired chamber to around 1500 to 1600 °C, the fluorine is driven off and the remaining calcium phosphate is of greater plant availability than the P in original rock phosphate.

The defluorination is done by two processes, namely (i) calcination, and (ii) fusion.

Calcination refers to the heating of rock phosphate with silica and steam to temperature below the melting point of the mixture. While, **fusion** refers to the process in which the mixture is heated to a temperature above its melting point so that the mixture is fused to form a glassy product. Regardless of whether fusion or calcination is employed, the reaction believed to take place during defluorination of rock phosphate with heat may be represented as follows:

$$\underset{}{Ca_{10}(PO_4)_6F_2} + x\,SiO_2 + \underset{\text{(Steam)}}{H_2O} \xrightarrow[\text{heating}]{(1500\text{ to }1600\ ^\circ C)} \underset{\text{(TCP)}}{3\,Ca_3(PO_4)_2} + \underset{\text{(Calcium silicate)}}{CaO.xSiO_2} + 2\,HF(\uparrow)$$

... (3.7)

or,

$$Ca_{10}(PO_4)_6F_2 + SiO_2 + \underset{\text{(Steam)}}{H_2O} \xrightarrow[\text{heating}]{1500 \text{ to } 1600\ ^\circ C} \underset{\text{(TCP)}}{3\,Ca_3(PO_4)_2} + \underset{\text{(Calcium silicate)}}{CaSiO_3} + 2\,HF(\uparrow) \quad \ldots (3.8)$$

The phosphorus remaining is α-tricalcium phosphate, and the 'F' from apatite escapes as 'HF' gas.

3.12. Production of Orthophosphoric Acid (H_3PO_4)

Orthophosphoric acid is a polybasic (tribasic) acid. It can be represented as shown below:

```
      O
      ||
HO – P – OH
      |
      OH
```

(H_3PO_4)

The OH^- group in H_3PO_4 is substituted by other group which may be called as primary, secondary or tertiary substitution. Thus, from H_3PO_4 acid, many products can be produced. So, it is an important intermediate product required for most of the P-fertilizers industry.

Phosphoric acid can be produced by two ways such as (i) **furnace process**, and (ii) **wet process**. In furnace process, electric-furnace is used extensively to make elemental P. The furnace process is also known as elemental P, most of which is converted to H_3PO_4 for non-fertilizer use. In case of wet process, some mineral acid is used to decompose rock phosphate. Usually, H_2SO_4, HCl and HNO_3 acids are used for acidulation. Based on the acid used, wet process may be classified according to the acid used. The share of production of wet process is 95% of that of total H_3PO_4 production, while the furnace process is only 5%. This is because the cost of wet process is much less compared to thermal or furnace process.

3.12.1. Furnace Process of H_3PO_4: Thermal Route

The first step of the furnace process of H_3PO_4 is to produce elemental P in an electrical furnace. Phosphate nodules or other phosphate materials, silica pebble and coke or coal are mixed and fed to the furnace. The electric current which enters the furnace through carbon or graphite electrode fuses the rock and silica, and the carbon in the coke or coal reduces the phosphate. A mixture of phosphorus vapour and CO gas is withdrawn continuously from the furnace. The P is then condensed to a liquid. Molten calcium silicate slag and an iron-phosphorus compound known as ferrophosphorus are tapped and removed from the furnace periodically. Elemental P thus produced is burnt in the air to get phosphorus pentaoxide (P_2O_5) in a separate plant, which in turn is reacted with water to get H_3PO_4. The principal reactions are represented by following equations:

$$Ca_{10}(PO_4)F_2 + 15\,C + 6\,SiO_2 \rightarrow 1.5\,P_4 + 15\,CO + 3(3CaO.2SiO_2) + CaF_2 \quad ...(3.9)$$

$$P_4 + 5\,O_2 \rightarrow 2\,P_2O_5 \quad ...(3.10)$$

$$P_2O_5 + 3\,H_2O \rightarrow 2\,H_3PO_4 \quad ...(3.11)$$

The "thermal grade acid" produced by this route is highly pure but very expensive and therefore rarely used for the production of P-fertilizers. Phosphoric acid made by thermal process is known as **white** or **furnace acid**. This method can be followed where electricity is cheaper (where hydro-electricity is available such as National Fertilizer Limited (NFL), Nangal, Punjab.

3.12.2. Wet Process of H_3PO_4: Acidulation Route

These processes involve disintegration of apatite with a mineral acid to release P as H_3PO_4. Wet processes are classified according to the acid used to decompose rock phosphate. Normally, HCl, HNO_3 and H_2SO_4 are used. Processes using HNO_3 lead to the production of nitrophosphates, while use of HCl acid is rather uneconomical and H_2SO_4 is the most common means of producing H_3PO_4 for fertilizer use. Phosphoric acid produced by treating rock phosphate with H_2SO_4 is known as **green** or **wet process acid**.

3.12.2.1. Principal reactions of H_3PO_4 production

Wet process H_3PO_4 acid is manufactured by extracting rock phosphate with H_2SO_4. The principal reaction taking place is represented by the following equation using pure fluorapatite rock phosphate:

$$Ca_{10}(PO_4)_6F_2 + 10\,H_2SO_4 + 20\,H_2O \rightarrow 10\,CaSO_4.2H_2O + 6\,H_3PO_4 + 2\,HF \quad ...(3.12)$$

or, $$Ca_{10}(PO_4)_6F_2 + 10\,H_2SO_4 + 10\,nH_2O \rightarrow 10\,CaSO_4.nH_2O + 6\,H_3PO_4 + 2\,HF \quad ...(3.13)$$

where, n = 0, 0.5 and 2.0, depending on the hydrate form in which calcium sulphate crystallizes.

The reaction in fact represents the net result of two stages. In the first stage, H_2SO_4 reacts with apatite forming monocalcium phosphate (MCP). In the second stage, MCP reacts with H_2SO_4 to form H_3PO_4 and $CaSO_4.nH_2O$. These two stages of reaction take place simultaneously in a single reactor.

The reaction is carried out for about 8 hours in a digestion chamber. The reaction is essentially complete within a matter of a few minutes. But the additional time is needed to ensure the formation of gypsum crystals of a size adequate to permit more rapid filtration. The slurry is ultimately filtered and the acid is concentrated to desired strength by evaporation.

It may be borne in mind that whenever apatite is taken, the classical fluorapatite [$Ca_{10}(PO_4)_6F_2$] formula is used as mentioned above. However, we must always keep in mind that this is not the only material which takes part in the reaction.

The simplified reactions with different acids may be represented by following equations:

$$Ca_{10}(PO_4)_6F_2 + HCl \rightarrow CaCl_2 + H_3PO_4 + HF \quad \dots (3.14)$$

$$Ca_{10}(PO_4)_6F_2 + HNO_3 \rightarrow Ca(NO_3)_2 + H_3PO_4 + HF \quad \dots (3.15)$$

$$Ca_{10}(PO_4)_6F_2 + H_2SO_4 \rightarrow CaSO_4 + H_3PO_4 + HF \quad \dots (3.16)$$

In all the 3 cases, H_3PO_4 is one of the desired products. Along with H_3PO_4, the respective salts are present, out of which $CaSO_4$ is sparingly soluble salt while other two salts, namely $CaCl_2$ and $Ca(NO_3)_2$ are very soluble in water. Thus, it is easier to separate $CaSO_4$ salt from H_3PO_4 acid but difficult to separate $CaCl_2$ and $Ca(NO_3)_2$ salt from H_3PO_4. By filtration we cannot separate $CaCl_2$ and $Ca(NO_3)_2$ salt from H_3PO_4. The $CaSO_4$, being a sparingly soluble salt, can be separated easily by filtration. However, $CaCl_2$ and $Ca(NO_3)_2$ can be removed by using some organic solvents like isoamyl alcohol, n-butyl alcohol or tertiary alcohol which can separate H_3PO_4, leaving other mixtures. So, by organic solvents $CaCl_2$ and $Ca(NO_3)_2$ can be separated, but it is costly. All commercial processes use H_2SO_4 for acidulation.

3.12.2.2. Fundamental Principles of H_3PO_4 Production

For production of H_3PO_4, following criteria are followed:

(1) **Selection of rock phosphate:** Very good quality rock phosphate is required. It should be rich in P_2O_5 content and impurities should be as low as possible (free from impurities). Powdered rock phosphate having 90% passes through 100-mesh sieve is used. The rock phosphate must have specifications of 31-35% P_2O_5, 45-49% CaO, 1.5-4% R_2O_3 and 2-8% SiO_2. Any materials which contain these specifications should be applicable in the production of H_3PO_4.

(2) **Choice of acid:** Any mineral acid can be used such as HCl, HNO_3 and H_2SO_4 or mixture of these acids can be used. However, H_2SO_4 is usually preferred over HCl and HNO_3. The H_2SO_4 having 93-98% concentration should be used.

(3) **Chemical reactions:** Primary reaction is represented as:

$$Ca_{10}(PO_4)_6F_2 + 10\,H_2SO_4 + 20\,H_2O = 10\,CaSO_4.2H_2O + 6\,H_3PO_4 + 2\,HF \quad ...(3.17)$$

It is the overall reaction. In reality, it is first converted to monocalcium phosphate by reacting apatite with H_2SO_4. The monocalcium phosphate is then reacts with H_2SO_4 to form H_3PO_4 and gypsum (gypsum). It should be noted that the form in which the $CaSO_4$ finally precipitates depends upon the reaction conditions. The $CaSO_4$ formed may exist in 3-forms, which is known as water of crystallization. It may be either in the form of (i) calcium sulphate anhydrite ($CaSO_4$); (ii) calcium sulphate semi-hydrate or hemihydrate [$CaSO_4.½H_2O$]; and (iii) calcium sulphate dihydrate (gypsum) [$CaSO_4.2H_2O$].

The important factor of these 3-formation is temperature. Anhydrite and hemihydrate forms are stable at very high temperature (Fig. 3.4). When temperature becomes low, gypsum is formed. So, finally gypsum is precipitated.

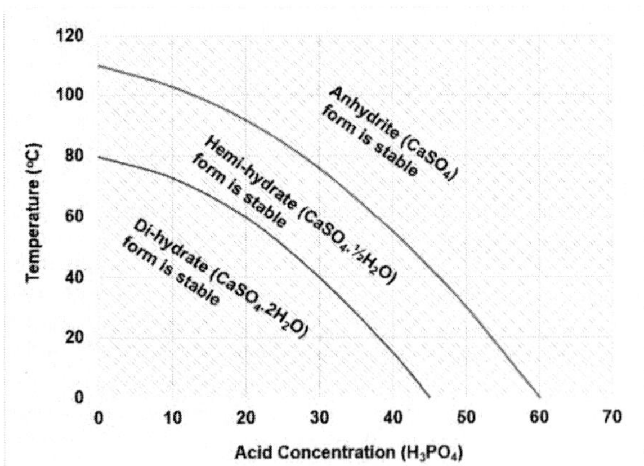

Fig. 3.4. Dependence of calcium sulphate dihydrate–hemihydrate equilibrium on temperature and P_2O_5 concentration

Concentration of H_3PO_4 and temperature are important. If temperature is 80 °C and acid concentration is 40%, then calcium sulphate dihydrate is formed which is stable. But, at high temperature (>100 °C) and acid concentration >60%, then first product is hemihydrate and finally anhydrite form of $CaSO_4$. The dihydrate process is most popular because they are relatively simple and adaptable to a wide range of grades and types of rock phosphates. Hemihydrate processes have the advantage of producing H_3PO_4 of a relatively high concentration. But the operating temperature is very high.

The dihydrate route involves following 4 essential steps:

(i) Disintegration/dissolution of rock phosphate with H_2SO_4 (3-4 minutes) and release of H_3PO_4.

(ii) Allowing the slurry of sufficiently long time (8-12 hours) to grow the gypsum crystals of adequate size.

(iii) Filtration/separation of H_3PO_4 from gypsum. The gypsum separated from H_3PO_4 is called phosphogypsum.

(iv) Concentration of product acid to commercial grade of acid. The H_3PO_4 acid produced by most of the dihydrate processes contains 26-32% P_2O_5. This is usually concentrated to 52-54% P_2O_5 by evaporation.

(4) Heat of Reaction: The reaction between rock phosphate and H_2SO_4 is exothermic *i.e.* heat is generated. It is estimated that about 400 Kcal/kg of P_2O_5 is liberated during the reaction. It may vary depending on the nature of rock phosphate and other impurities present in it. This actual heat depends on types of rock phosphate, concentration of acid, *etc.*

(5) Time of Reaction: The time required to complete the reaction may depends on two factors.

(a) Actual time required for neutralizing the apatite with H_2SO_4. This is about 2-3 minutes only. That means disintegration of rock phosphate with acid requires 3 minutes.

(b) Time required for crystallization of gypsum takes about 1.5 to 12 hours. The growth of crystal is a physical reaction.

Thus, there is a wide difference or range of time required for complete reaction. The actual time required is fixed on the basis of what type of technology or which route of H_3PO_4 production is adopted. Once the parameter is fixed, the reaction or the procedure is mechanical.

3.12.2.3. Other Reactions Occur during H_3PO_4 Production

Minor reactions arising from impurities in the rock phosphate are of considerable importance in plant operation. Since rock phosphate contains other impurities beside apatite. So, H_2SO_4 will also react with those impurities. The actual amount of H_2SO_4 needed is much more than theoretical values of H_2SO_4. Depending on the amount of impurities, the quantity of H_2SO_4 will vary.

(i) Reaction with Hydrofluoric Acid (HF) and Silica

Reaction between fluoride (F) present in rock phosphate and H_2SO_4 can be represented as:

$$CaF_2 + H_2SO_4 = CaSO_4 + 2\,HF \qquad \text{...(3.18)}$$

Hydrofluoric acid (HF) is highly reactive. The liberated HF reacts with silicates or silica (SiO_2) present in rock phosphate to form hydrofluosilicic acid (H_2SiF_6).

$$6\,HF + SiO_2 = \underset{\text{(Hydrofluosilicic acid)}}{H_2SiF_6} + 2\,H_2O \qquad \text{...(3.19)}$$

(ii) Reactions with Na and K-Salts

Hydrofluosilicic acid is less reactive than hydrofluoric acid. But it can react with Na and K-salts present in the rock phosphate to form a corresponding fluosilicate salts.

$$Na_2O + H_2SiF_6 = \underset{\text{(Na-fluosilicate)}}{Na_2SiF_6} + H_2O \qquad \text{...(3.20)}$$

$$K_2O + H_2SiF_6 = \underset{\text{(K-fluosilicate)}}{K_2SiF_6} + H_2O \qquad \text{...(3.21)}$$

These fluosilicates have steep temperature to solubility curves, and precipitate in pipelines, filter cloths and form a very hard salt.

At high temperature and acid conditions of the process, excess fluosilicic acid is decomposed to form gaseous silicon tetrafluoride (SiF_4) and hydrofluoric acid (HF).

$$H_2SiF_6 + \text{Heat} + \text{Acid} \rightarrow \underset{\text{(Silicon tetrafluoride)}}{SiF_4} + 2HF \qquad \text{...(3.22)}$$

The silica is invariably precipitated in the gas ducts and scrubbing systems of the plant as a gelatinous solid which will cause blockages unless facilities for removal are provided.

All these compounds of HF, H_2SiF_6 and SiF_4 are hazardous. Therefore, these compounds should be collected and removed safely.

(iii) Reaction of H_3PO_4 with Sesquioxide

$$2\,H_3PO_4 + R_2O_3 = 2\,RPO_4 + 3\,H_2O \text{ (where, R = Al or Fe)} \qquad \text{...(3.23)}$$

$$2\,H_3PO_4 + Fe_2O_3 = 2\,FePO_4 + 3\,H_2O \qquad \text{...(3.24)}$$

$$2\,H_3PO_4 + Al_2O_3 = 2\,AlPO_4 + 3\,H_2O \qquad \text{...(3.25)}$$

Ferric oxide and similar impurities in the rock phosphate reacts with H_3PO_4 to form corresponding phosphates. This particular reaction implies three things:

(a) A part of H_3PO_4 is lost.

(b) $AlPO_4$ or $FePO_4$ form as sludge which interfere in the product. Because of the presence of this sludge it gives green colour (green H_3PO_4).

(c) They can also reduce the water soluble P_2O_5 content of a high-grade fertilizer.

(iv) Reaction of H_2SO_4 with $CaCO_3$

Calcium carbonate ($CaCO_3$) and magnesium carbonate ($MgCO_3$) present in the rock phosphate may react with H_2SO_4 to form gypsum and CO_2 before reacting with apatite.

$$H_2SO_4 + CaCO_3 = CaSO_4 + H_2O + CO_2 \quad \ldots (3.26)$$

or, $$H_2SO_4 + CaCO_3 + H_2O = CaSO_4.2H_2O + CO_2 \quad \ldots (3.27)$$

This reaction implies that some amount of H_2SO_4 is lost or wasted. Thus, H_2SO_4 requirement increases with the amount of carbonates in rock phosphate. Therefore, cost of production goes up. Moreover, CO_2 evolved may forms foam or froth. This is a nuisance of this reaction.

(v) Reaction with Chloride

Chloride comes from rock phosphate as well as water used for processing. Chloride in presence of H_2SO_4 and high temperature becomes highly corrosive and the equipments are damaged.

3.12.2.4. Wet Process Phosphoric Acid Production Technology

In the wet process of phosphoric acid (H_3PO_4) production, rock phosphate and H_2SO_4 are passed in a reaction tank (reactor) and mixed thoroughly as shown schematically in Fig. 3.5. The slurry is formed in the reactor. Due to the exothermic reaction, heat is generated which goes out through the vent. Other hazardous gases also go out through the vent. Cool air is then passed to the reactor for cooling the slurry.

The gypsum ($CaSO_4.2H_2O$) is formed as slurry. This slurry is then passed through filter feed tank. The material is passed through nylon pan or nylon cloth (1.5 to 2″). Then suction is created. The product H_3PO_4 is then collected which contains 26 to 32% P_2O_5.

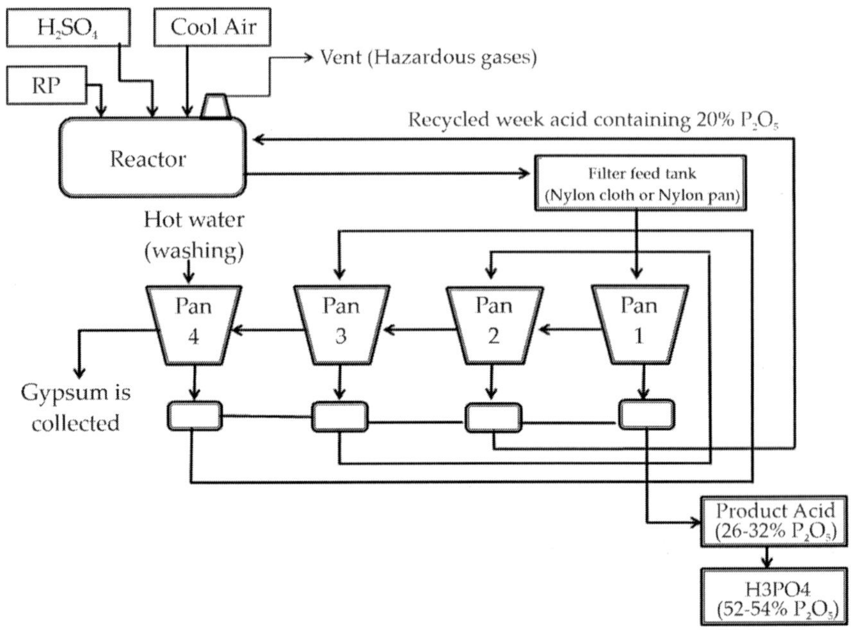

Fig. 3.5. Schematic diagram of wet process of phosphoric acid production

Some amount of acid is still remained in the tank. Thus, continuous washing is done for several times. For this, in 4th pan, hot water is passed and thus a cyclic path is followed. Washed material is thus recycled. The product acid containing 26 to 32% P_2O_5 is then taken in a suitable vessel and the material is concentrated to 52-54% P_2O_5. This is done by passing hot air to the diluted acid (26 to 32% P_2O_5). Here, water is evaporated and acid becomes concentrated. The gypsum ($CaSO_4.2H_2O$) thus produced is known as phosphogypsum which contain 2% P_2O_5 and used as an amendment. The concentrated H_3PO_4 containing 52-54% P_2O_5 can be transported in special tankers and marketed. Through pipeline also it is transported. This is used for many purposes such as production of superphosphoric acid or polyphosphoric acid, TSP and other fertilizers.

3.13. Super Phosphoric Acid or Polyphosphoric Acid

Tennessee Valley Authority (TVA) have developed a method of producing highly concentrated acid, usually referred to as "**super phosphoric acid**", that contains over half of its P in the pyro and more highly condensed forms.

When made by the thermal process, H_3PO_4 acid contains 33-37% P (76-85% P_2O_5) which is equivalent to 106-115% H_3PO_4. It is produced by limiting the amount of water and providing extra cooling. The product is fluid at ordinary temperature because it has strong super cooling tendencies. Super phosphoric acid is produced to reduce the storage, handling and transportation costs of phosphoric acid.

3.13.1 Production Technology of Super Phosphoric Acid

Wet process super phosphoric is made by evaporation of water to concentrate H_3PO_4 beyond the equivalent of 100% H_3PO_4.

The P_2O_5 when reacts with H_2O form several types of phosphoric acid as mentioned below:

$$P_2O_5 + H_2O \rightarrow 2\,HPO_3 \quad \text{(Metaphosphoric acid, MPA)} \qquad \ldots (3.28)$$

$$P_2O_5 + 2H_2O \rightarrow H_4P_2O_7 \quad \text{(Pyrophosphoric acid, PPA)} \qquad \ldots (3.29)$$

$$P_2O_5 + 3H_2O \rightarrow 2\,H_3PO_4 \quad \text{(Orthophosphoric acid, OPA)} \qquad \ldots (3.30)$$

$$P_2O_5 + nH_2O \rightarrow H_{(n+2)}P_nO_{(3n+1)} \quad \text{(Polyphosphoric acid or super phosphoric acid)} \qquad \ldots (3.31)$$

When H_2O is removed from H_3PO_4, we will get the reverse reactions.

$$\underset{\text{(OPA)}}{2\,H_3PO_4} \xrightarrow{-H_2O} \underset{\text{(PPA)}}{H_4P_2O_7} \xrightarrow{-H_2O} \underset{\text{(MPA)}}{2\,HPO_3} \qquad \ldots (3.32)$$

Thus, by removing more and more water from OPA (H_3PO_4), meta- and phyro-form of acid is produced. But whole amount of H_3PO_4 does not convert to pyro- and meta-form.

Further, one mole of pyrophosphoric acid ($H_4P_2O_7$) reacts with one mole of metaphosphoric acid (HPO_3) to form $H_5P_3O_{10}$ acid (Tri-polyphosphoric acid, Tri PPA) which is an unstable form.

$$\underset{\text{(PPA)}}{H_4P_2O_7} + \underset{\text{(MPA)}}{HPO_3} \rightarrow \underset{\text{(Tri-TPA)}}{H_5P_3O_{10}} \qquad \ldots (3.33)$$

Again tri-polyphosphoric acid combines with another mole of HPO_3 to form tetra-polyphosphoric acid (Tetra-PPA).

$$\underset{\text{(Tri-PPA)}}{H_5P_3O_{10}} + \underset{\text{(MPA)}}{HPO_3} \rightarrow \underset{\text{(Tetra-PPA)}}{H_6P_4O_{13}} \qquad \ldots (3.34)$$

So, starting from tri- (P3) to tetra- (P4) up to deca-polyphosphate (P_{10}) (Deca-PPA) may be obtained. These different polyphosphates may be represented with general formula of polymers $H_{n+2}P_nO_{3n+1}$. At the end, we get some amount of ortho-, some pyro-, some meta- and varying amount of tri- (P3) to deca- (P_{10}) forms of polymers. All these polymers are present in equilibrium with tri-polyphosphate to deca-polyphosphate are in decreasing amounts.

All these polymers can also be expressed in following simplified ways:

$H_4P_2O_7$ Pyrophosphoric acid
$+ HPO_3$ Metaphosphoric acid (MPA)

$H_5P_3O_{10}$ Tri-polyphosphoric acid (Tri-PPA)
$+ HPO_3$ Metaphosphoric acid (MPA)

$H_6P_4O_{13}$ Tetra-polyphosphoric acid (Tetra-PPA)
$+ HPO_3$ Metaphosphoric acid (MPA)

$H_7P_5O_{16}$ Penta-polyphosphoric acid (Penta-PPA)
$+ HPO_3$ Metaphosphoric acid (MPA)

$H_8P_6O_{19}$ Hexa-polyphosphoric acid (Hexa-PPA)
$+ HPO_3$ Metaphosphoric acid (MPA)

$H_9P_7O_{22}$ Hepta-polyphosphoric acid (Hepta-PPA)
$+ HPO_3$ Metaphosphoric acid (MPA)

$H_{10}P_8O_{25}$ Octa-polyphosphoric acid (Octa-PPA)
$+ HPO_3$ Metaphosphoric acid (MPA)

$H_{11}P_9O_{28}$ Nona-polyphosphoric acid (Nona-PPA)
$+ HPO_3$ Metaphosphoric acid (MPA)

$H_{12}P_{10}O_{31}$ Deca-polyphosphoric acid (Deca-PPA)

....

....

$H_4P_2O_7$ Pyrophosphoric acid
$+ n.HPO_3$ n-moles of metaphosphoric acid

$H_{(n+4)}P_{(n+2)}O_{(3n+7)}$ (Poly-phosphoric acid)

If we have a polyphosphoric acid containing 76% P_2O_5, it will have 40% H_3PO_4 (ortho-phosphoric acid), 40% $H_4P_2O_7$ (pyro-phosphoric acid), 20% higher forms (Tri-, tetra-, penta-, hexa-, hepta-, octa-, nona- and deca-polyphosphoric acid). At 85% P_2O_5 polyphosphoric acid, it may contain 10-20% each of ortho-, pyro-, tri-, tetra-, penta-, hexa-, hepta-, octa-, nona- and deca-phosphoric acid. Therefore, with higher concentration, relative proportion of ortho- and pyro- comes down and concentration of tri-, tetra-, penta-, hexa-, hepta-, octa-, nona- and deca- (polymers) goes up. Agricultural production of polyphosphoric acid contains 69-75% P_2O_5 usually.

Besides its much higher P_2O_5 content and the savings in the handling, storage and transportation costs, super phosphoric acid contains about 50% of its P_2O_5 in the form of polyphosphoric acids whereas, acid of ordinary concentration is essentially all orthophosphoric acid.

3.13.2. Advantages of Polyphosphoric acids

The advantages of polyphosphoric acids are:

(i) They are high analysis fertilizer.

(ii) It can be used for production of liquid fertilizer.

(iii) It can be used for production of ammoniated super phosphate (Ammonium polyphosphate).

$H_3PO_4 + NH_3 \rightarrow NH_4H_2PO_4$ (Ammonium orthophosphate) ... (3.35)

$H_4P_2O_7 + NH_3 \rightarrow NH_4H_3P_2O_7$ (Ammonium pyrophosphate) ... (3.36)

$H_{n+2}P_nO_{3n+1} + NH_3 \rightarrow NH_4H_{n+1}P_nO_{3n+1}$ ... (3.37)
(Polyphosphoric acid) (Ammonium polyphosphate)

3.13.3. Fate of Polyphosphates in soils

Polyphosphoric acid, being a mixture of acids when it is applied to soil, it undergoes hydrolysis and form H_3PO_4 (ortho-phosphoric acid) which is available to the plant. Other forms of polyphosphoric acid are not used by the plants as such unless they undergo hydrolysis to form H_3PO_4.

When we apply polyphosphate in soil, since it contains orthophosphate form, plant takes this either $H_2PO_4^-$ or HPO_4^{2-} forms (both are orthophosphate ions). But other higher forms of phosphate are not taken by plants. Then, what is the fate of them? In fact, $H_4P_2O_7$ or other forms undergo hydrolysis in soil and ultimately forms ortho-forms as shown below:

$$H_4P_2O_7 + H_2O \xrightarrow{\text{hydrolysis}} 2\,H_3PO_4 \quad \text{... (3.38)}$$

(This is then used by the plant)

This hydrolysis is a rapid process. But if soil is acidic, the rate of hydrolysis may be very slow. It takes more time to form ortho-forms. Thus, polyphosphate is advantageous where soil is acidic and more P-fixation is expected to occur because of more sesquioxide (R_2O_3) over fully orthophosphoric form as in case of monocalcium phosphate [MCP, $Ca(H_2PO_4)_2.H_2O$].

3.13.4. Chelated Polyphosphate

Polyphosphate may form chelated compound with micronutrients (Zn, Cu and Mn). When polyphosphate and micronutrients are applied separately it is very difficult because micronutrients are required by the plant at a very small quantity. Therefore, uniform distribution in the field is very difficult.

When micronutrients are mixed with polyphosphates, they react with micronutrients and keep them in solution form. This solution of polyphosphate and micronutrients is ammoniated with NH_3 and thus, chelated ammonium phosphate is formed. Here distribution of micronutrients is uniform.

3.14. Production of Super Phosphate

In 1842, **John Bennet Lawes**, produced a fertilizer called "**Superphosphate**" in England and took a patent for this in 1843. He actually produced superphosphate by treating coprolites (P-bearing fossilized dung) with H_2SO_4.

$$\underset{\text{(P-bearing fossilized dung)}}{\text{Coprolite}} + H_2SO_4 \rightarrow \text{Superphosphate} \quad \ldots (3.39)$$

This may be considered as the first commercially successful venture of fertilizer production. Later on people use bones instead of coprolite. Then rock phosphate has come for commercial production.

In India, in 1906, single superphosphate (SSP) is produced for the first time by EID-Parry (India) Ltd, at Ranipet (about 120 km away from Chennai). They used bones and H_2SO_4 as the raw materials to produce SSP. Subsequently, they used rock phosphate in place of bones. In fact most of the countries produced SSP as the first fertilizer, because it is a simple process and raw materials are easily available.

3.14.1. Definition of Superphosphate

The name "Superphosphate" applies to fertilizer obtained by treating finely ground rock phosphate with H_2SO_4, H_3PO_4 or a mixture of H_2SO_4 + H_3PO_4.

Therefore, superphosphate refers a group of materials which are produced by reacting rock phosphate with H_2SO_4, H_3PO_4 or H_2SO_4 + H_3PO_4.

Rock phosphate + $H_2SO_4 \rightarrow$ Single superphosphate (SSP) ... (3.40)

Rock phosphate + $H_3PO_4 \rightarrow$ Triple superphosphate (TSP) ... (3.41)

Rock phosphate + ($H_2SO_4 + H_3PO_4$) $\rightarrow$ Double superphosphate (DSP) ... (3.42)

The product made with H_2SO_4 usually contains 16-21% P_2O_5 and it is referred to as "**ordinary**" or "**normal superphosphate**" or simply "**superphosphate**". The product made with H_3PO_4 usually contains 43-48% P_2O_5 and is referred to as "**concentrated**" or "**triple superphosphate**".

The TSP is first produced in 1878 in Germany. But it was not popularized because it produced a product having only 8-15% P_2O_5. Later on, TSP was produced by others. However, it becomes popular only after 1930 because of its economic reason (it is a high analysis fertilizer).

Double superphosphate (DSP) is also known as enriched superphosphate. It is essentially a mixture of SSP and TSP. This is produced by reacting rock phosphate with a mixed acid system consisting of variable proportion of H_2SO_4 and H_3PO_4. The advantage of this material is a fairly high P_2O_5 content (25-35%) and the presence of some amounts of sulphates and calcium. It is not produced now-a-days anywhere in the world. It is of historic importance.

The production technology of both SSP and TSP are similar. Both hold an important place in the fertilizer industry primarily because they are cheap to produce. But, at present the contribution of SSP to world phosphate supply is shrinking and its present share among phosphate fertilizers is not more than 20% in India.

3.14.2. Advantages of Single Superphosphate

Single superphosphate (SSP) although contains 16% P_2O_5 (considered as low analysis fertilizer), it is an important P-fertilizer and is likely to remain so because of the following advantages:

(i) The production process is very simple, requiring little technological skill.

(ii) Besides P, it supplies two secondary nutrients namely, calcium (Ca) and sulphur (S).

(iii) It is a multi-nutrient fertilizer.

(iv) The fertilizer effectiveness of SSP is unquestioned.

(v) Cost of installation of plant is very low, thus small plant can also be viable.

Its major disadvantage is of low analysis (16% P_2O_5), consequently high cost of storage, transportation and application per unit of P_2O_5. Thus, SSP is not given emphasis because of low analysis. Over a period of time, it has lost its position as many fertilizers come in the market. Presently, more than 80% of total P comes from complex NP/NPK fertilizer.

3.14.3. Production Technology of Single Superphosphate

The production of normal superphosphate or single superphosphate (SSP) is basically a simple operation. The only ingredients required are ground rock phosphate and H_2SO_4. The main chemical reaction that occurs when finely ground rock phosphate is mixed with H_2SO_4 in the manufacture of SSP may be represented by the following equation:

$$Ca_{10}(PO_4)_6F_2 + 7\,H_2SO_4 + 17\,H_2O \rightarrow \underset{(MCP)}{3\,Ca(H_2PO_4)_2.H_2O} + 7\,CaSO_4.2H_2O + 2\,HF(\uparrow) \quad \ldots (3.43)$$

It is generally agreed that the reactions proceed in two-stages:

(a) Sulphuric acid reacts with a part of apatite to form H_3PO_4, and

(b) The resultant H_3PO_4 subsequently reacts with the remaining apatite forming monocalcium phosphate (MCP).

While the first step is rapid, the second step proceeds slowly. Thus, phosphate originally present as apatite is converted to MCP [$Ca(H_2PO_4)_2.H_2O$] which is water soluble form. Calcium sulphate (gypsum) is also formed and is inter mixed with MCP, the former ($CaSO_4$) being the dominant member. This is the reason for its low analysis. In SSP, the ratio of MCP: gypsum is 3:7. That means 70% of the material is in the form of gypsum and only 30% is MCP.

The production of SSP involves the following operations:

(i) Mixing of Rock Phosphate with Acid

The powered rock phosphate containing 30-38% P_2O_5 is mixed with H_2SO_4 to form a slurry. Usually, 100 kg rock phosphate containing 30% P_2O_5 require 60 kg of H_2SO_4 (conc. 98%, sp. gr. 1.84 g cm^{-3}) for 100% acidulation. It takes about 5-10 minutes.

The mixing can be done either in **batch method** or **continuous method**. For many years, the only method of making SSP was to mix rock phosphate and H_2SO_4 in batches, known as batch method. In batch method, rock phosphate and H_2SO_4 are charged to a pan mixer, mixed for 1-2 minutes and discharged in the form of a slurry or paste into a bin or den.

Modern plants utilize a continuous process, in which a stream of finely ground rock phosphate is reacted with a stream of H_2SO_4. In the usual arrangement, H_2SO_4 and pulverized rock phosphate are metered continuously to a cone mixer. The mixer is a funnel-shaped device in which jets of acid and water are injected. The flow diagram of manufacturing process for superphosphate is shown in Fig. 3.6.

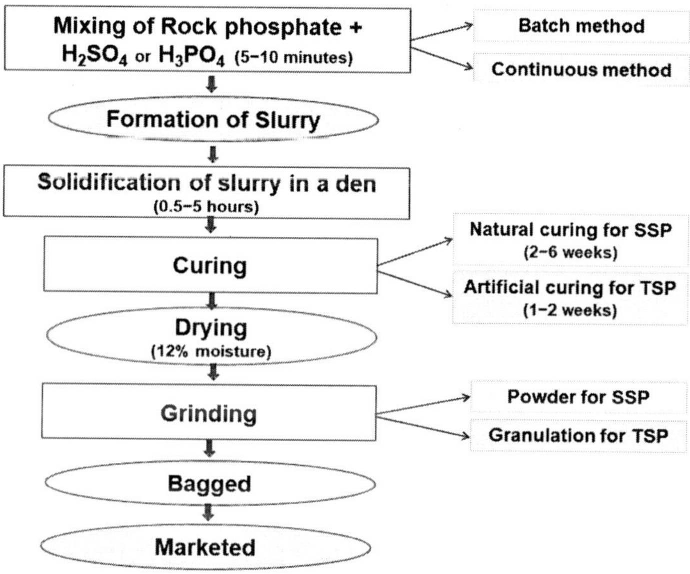

Fig. 3.6. Flow diagram of manufacturing process for superphosphate

(ii) Denning

Solidification of the slurry in a **den** (denning), where continuous reactions and crystallization of MCP take place (time required is 0.5 to 4 hours). The 'Den' is a long rectangular bin fitted tightly over or slow moving slat conveyor. The dens are of various design and may hold 100 tonnes or more materials. The acidulated material, fed into one end of the den, quickly solidifies and is moved as a continuous block to the particles by a rotary cutter. The reaction time in the den is 30 minutes to 4 hours. Curing and drying of the acidulated material is done in a den which is called as the "**heart**" of the plant. After mixing of rock phosphate and acid (H_2SO_4 or H_3PO_4), a semi-solid mixture is produced. This is collected in a den. It is nothing but the tank, where gaseous product is excavated and removed. Den has the capacity to collect the hazardous gases and collect the slurry. It allows the slurry to keep during the period of time required (0.5 to 4 hours). It converts or allows the slurry to form hard mass. After 6-24 hours in the den, the SSP, now a friable solid mass, is removed to curing piles for several weeks to allow complete chemical reactions.

(iii) Curing

The product is then passed to a storage pile or curing pile for curing. The wet material is subjected to final curing, when some of the slow reactions reach a state of near completion and the excess water evaporates. The curing time is 2-6 weeks, depending on the nature and proportion of raw materials and the conditions of manufacturer. The dried material is crushed into a coarse powder before it is distributed as SSP. Most of the cases, moisture is allowed to 12%. No other fertilizer is allowed such amount of moisture in the final product. The objective of curing is to allow sufficient time to come in equilibrium and complete the reactions. It also allows the material to dry. Excess water evaporates during this period. Curing is done by naturally or artificially. Curing in natural condition is done in case of SSP and it requires 2-6 weeks. While, artificial drying is done in case of TSP (1-2 weeks). Since SSP is a low analysis fertilizer, the cost of drying by artificial method is more than naturally drying production of SSP. But since TSP is high analysis fertilizer, artificial drying is followed for TSP.

(iv) Grinding

The cured material is excavated from the pile and pulverized/granulated before bagging and marketing. If granular SSP is desired, it can be granulated either before or after it is cured.

3.14.4. Production Technology of Triple Superphosphate or Concentrated Phosphate

Production technology of triple superphosphate (TSP) or concentrated phosphate is more or less similar to that of SSP. The TSP is produced by reacting rock phosphate with H_3PO_4 as shown below:

$$Ca_{10}(PO_4)_6F_2 + 14\,H_3PO_4 + 10\,H_2O \rightarrow 10\,Ca(H_2PO_4)_2.H_2O + 2\,HF \qquad \ldots (3.44)$$

The TSP is a highly concentrated P-fertilizer containing P in the form of water soluble P (WSP) *i.e.* monocalcium phosphate (MCP) which is water soluble. Another advantage is that a part of its P comes directly from H_3PO_4. The TSP may be produced in either granular form or non-granular form with 46% total P_2O_5 out of which 42.5% WSP and remaining 3.5% is citrate soluble P (CSP).

Some of the disadvantages of TSP are its acidic character, which may cause deterioration of bags made up of jute/hemp or paper and its unsuitability for blending with urea. The main phosphate compound present in TSP is MCP which is the same as in case of SSP. However, TSP consists almost entirely of MCP, whereas SSP contains a large proportion of gypsum ($CaSO_4.2H_2O$).

3.14.5. Steps of Triple Superphosphate Production

(i) **Mixing of rock phosphate with acid:** In the manufacture of TSP, the rock phosphate and acid (H_3PO_4) are mixed in a mixing chamber. Mixing can be done by batches or continuous mixing.

(a) **Batches (Batch mixing):** Known quantity of rock phosphate is mixed with required amount of acid and mechanical mixing is done. After mixing, whole amount of slurry is taken out and next batch is taken.

(b) **Continuous mixing:** Here hopper (cone shaped container) is there in which rock phosphate at a pre-determined rate is allowed to fall and acid is allowed to fall simultaneously. Here, continuous mixing is done. No gap as in case of batch mixing.

Choice of rock phosphate: The rock phosphate particle size generally is same as in case of SSP and TSP *i.e.* 90% of material should pass through 100-mesh sieve while, the acid concentration usually is 75% H_3PO_4. Selection of rock phosphate is very important. If we want to produce a product containing 16% P_2O_5 (as in SSP) we must have a rock phosphate having 30% P_2O_5. Similarly, for producing 20% P_2O_5 material it requires RP of 33% P_2O_5.

Depending on the rock phosphate of different grades, one can produce his desire grade of SSP by mixing rock phosphate with acid so that the ultimate product may be the local desire grade. Apart from P_2O_5, the rock phosphate should have impurities as low as possible. More impurities means higher consumption of acids and more cost of final product.

(ii) **Den:** The acidulated product is denned as in case of SSP. It is a place where slurry is taken. It is the **heart** of any superphosphate industry. The functions of den are:

(a) To receive and retain the slurry without allowing slurry to flow here and there;

(b) There should be enough provision for collecting all obnoxious gases produced from the reactions. They are safely disposed off; and

(c) During cooling of the product in den, the material becomes solid by natural process. So den should be constructed in such a way that solidified material is easily removed or excavated.

(iii) **Curing:** The solidified material is cured in curing storage. The TSP solidifies much more rapidly than SSP and the two operations differ mainly in that respect.

(iv) **Granulation:** The cured material is excavated from the pile as in case of SSP and then granulated using a granulator or spherodizer.

3.14.6. Specification of Single Superphosphate and Triple Superphosphate

Some regulations as specified by Fertiliser Control (Order) 1985 and Essential Commodities Act. 1956 (FCO, 1985) are given in Table 3.7.

Table 3.7. Specification of single superphosphate (SSP) and triple superphosphate (TSP)

Specifications	SSP	TSP
Moisture content (max.)	12%	12%
Free phosphoric acid (max.)	4% H_3PO_4 (as P_2O_5)	3% H_3PO_4 (as P_2O_5)
Water soluble P (min.)	16% P_2O_5	42.5% P_2O_5
Total P	16% P_2O_5	46% P_2O_5

Source: FCO (2019)

In case of SSP, the product containing about 16-22% P_2O_5 (WSP) is produced in all over the world, while, in case of TSP, the product contains about 40-49% P_2O_5 (WSP) in all over the world. In India, the TSP produced contains 42.5% P_2O_5 (WSP). For 22% P_2O_5 product, rock phosphate should have minimum impurities.

3.14.7. Comparison between SSP and TSP production technology

A brief comparison between single superphosphate (SSP) and triple superphosphate (TSP) production technology are given in Table 3.8.

3.15. Ammoniation of Superphosphate

Ammoniation of superphosphate refers to addition of ammonia (NH_3) in superphosphate. When superphosphate is neutralized by reacting anhydrous

Table 3.8. Comparison between single superphosphate (SSP) and triple superphosphate (TSP) production technology

Parameter	SSP	TSP
Acid	• Theoretically, 60 kg H_2SO_4 (98% conc.) is required for 100 kg RP (30% P_2O_5) to produce SSP. • Usual concentration of H_2SO_4 is 70%. The commercial grade H_2SO_4 (93-98%) can be pre-diluted or it can be simultaneously added during mixing.	• 0.82 kg H_3PO_4 is mixed with 1.35 kg RP (33% P_2O_5) for 1 kg P_2O_5 in TSP. • Usually, H_3PO_4 having 52-54% P_2O_5 is used without dilution of commercial one. • But we can use 25-50% P_2O_5 grade H_3PO_4.
Reactions	$Ca_{10}(PO_4)_6F_2 + 7\ H_2SO_4 + 17\ H_2O \rightarrow 3\ Ca(H_2PO_4)_2.H_2O +$ (MCP) $7\ CaSO_4.2H_2O + 2\ HF$ (gypsum) • For production of H_3PO_4 (OPA) we require 10 H_2SO_4, but for SSP it is 7 H_2SO_4. • $Ca_{10}(PO_4)_6F_2 + 10\ H_2SO_4 + 20\ H_2O \rightarrow 10\ CaSO_4.2H_2O + 6\ H_3PO_4 + 2\ HF$ • So, basically OPA and SSP preparation is same except the main product. • For SSP, each 3 moles of MCP, 7 moles of gypsum is produced. So, gypsum is more and P-containing compound (MCP) is less. • SSP always supply some quantity of other nutrients like Ca and S.	$Ca_{10}(PO_4)_6F_2 + 14\ H_3PO_4 + 10\ H_2O \rightarrow 10\ Ca(H_2PO_4).H_2O + 2\ HF$ (MCP) • Whole apatite is converted to MCP, no gypsum formation. • Some amount of P comes from H_3PO_4 itself. • In both the reactions, it is the empirical expression. • In reality some other reactions are also involved which depends on the impurities present in rock phosphate. So actual amount of acid is calculated by error and trail method. • Other hazardous gases like HF, SiF_4 and H_2SiF_6 are also liberated during both reactions. • TSP always supply Ca but not S.
Curing	• For economic reason, SSP is cured is natural condition because it is a low analysis, otherwise cost will increase.	• TSP is seldom cured under natural condition and usually artificial drying is done for TSP.
Grinding/ Granulation	• SSP is powdered and marketed. • Bagged in polythene bag as powdered material cannot be bagged in jute bag.	• TSP is granulated not powdered as in SSP and marketed. • Advantage of granulated materials over powder materials are: (i) Can be bagged in jute or polythene bags. If powder, it cannot be bagged in jute bag. Some losses will be there. (ii) Application of granular fertilizer is easy. Powder material during windy day is difficult to apply. (iii) Granular product can be used in seed-cum fertilizer drill. • Bagged in jute or polythene bag.

NH_3, ammonium superphosphate is formed. The cost of production of liquid ammonia (anhydrous NH_3) is less compared to any solid N-fertilizers like $(NH_4)_2SO_4$, NH_4NO_3, NH_4Cl or urea. Anhydrous NH_3 is alkaline is nature. On the other hand superphosphate either SSP or TSP contains some amount of free acids (H_3PO_4) in addition to MCP [$Ca(H_2PO_4).H_2O$]. The TSP contains 3% H_3PO_4 (as P_2O_5). While, SSP contains 4% H_3PO_4 (as P_2O_5). Moreover, MCP [$Ca(H_2PO_4).H_2O$] is acidic, 1 *N* solution of MCP is having pH 4.0. Therefore, when NH_3 comes in contact with SSP or TSP, it produces some other compounds leading to form a stable neutral compound.

So, NH_3 successfully react with H_3PO_4 or $Ca(H_2PO_4)_2$ as shown below:

$$H_3PO_4 + NH_3 \rightarrow NH_4H_2PO_4 \quad \ldots (3.45)$$

(Monoammonium phosphate, MAP) (Acidic)

$$H_3PO_4 + 2\,NH_3 \rightarrow (NH_4)_2HPO_4 \quad \ldots (3.46)$$

(Diammonium phosphate, DAP) stable (Neural)

$$H_3PO_4 + 3\,NH_3 \rightarrow (NH_4)_3PO_4 \quad \ldots (3.47)$$

(Triammonium phosphate, TAP) (Alkaline)

The TAP is unstable and decomposes because of high concentration of NH_3 (auto decomposition). So, depending on the amount of NH_3, MAP, DAP or TAP is produced. Therefore, NH_3 content should be added in such way that either MAP or DAP are formed.

Similarly, when MCP reacts with NH_3, it forms DCP ($CaHPO_4$) and MAP ($NH_4H_2PO_4$) as:

$$\underset{\text{(MCP)}}{Ca(H_2PO_4)_2} + NH_3 \rightarrow \underset{\text{(DCP)}}{CaHPO_4} + \underset{\text{(MAP)}}{NH_4H_2PO_4} \quad \ldots (3.48)$$

The above reaction is a double decomposition reaction. The MCP is water soluble phosphate, while DCP is water insoluble but citrate soluble phosphate, thus DCP is plant available P. So, when MCP reacts with limited amount of NH_3, there is no loss of availability of plant nutrients.

When larger quantity of NH_3 is used or ammoniation is done continuously, then TCP and TAP are formed as follows:

$$3\,\underset{\text{(MCP)}}{Ca(H_2PO_4)_2} + 12\,NH_3 \rightarrow \underset{\text{(TCP)}}{Ca_3(PO_4)_2} + 4\,\underset{\text{(TAP)}}{(NH_4)_3PO_4} \quad \ldots (3.49)$$

Since, TCP is water insoluble as well as citrate insoluble form, the solubility of phosphate decreases. So, excessive ammoniation should be stopped.

3.15.1. Advantages and Disadvantages of Ammoniation of Superphosphate

(i) The most important advantage of ammoniation of superphosphate is that through this process some amount of N can be introduced in the final product. Ammonia is the initial compound produced during the manufacture of synthetic N-products, and is less costly to produce than other compounds (products) requiring further processing.

(ii) It is fortunate that NH_3 reacts readily with superphoshate to form stable compounds. Thus, if we can prepare a fertilizer containing NH_3, the cost of production of fertilizer is cheap than any N-fertilizer like urea, $(NH_4)_2SO_4$ for which cost of production is high.

(iii) When ammonification of superphosphate is done, the cost of production is cheap. Transport cost is also cheap. It is also neutral compound, thus not hazardous as compared to NH_3 as such.

(iv) The disadvantage of ammoniation of superphosphate is that some amount of water soluble P is lost or reduced in this process.

(v) If the degree of ammoniation is more, the water soluble P is irreversibly lost as TCP or other forms.

Why ammoniation is done?

(i) It is done with view to supply some N at a very cheaper cost. The NH_3 is cheap itself (cost of production), but cost of storage, transport and application is more. If we incorporate NH_3 with other fertilizer (say superphosphate), then cost will be less.

(ii) So, basic concept is to supply some amount of N with reasonable cost.

3.15.2. Factors affecting ammonication

Ammoniation of superphosphate depends on the following factors:

(1) **Particle size of superphosphate:** Ammoniation is done only to powdered material. Finer is the particle size, more is the surface area. More is the surface area, more will be the reaction with NH_3. However, very fine powder is not desirable. It will be too dusty for ammoniation. Normally, amount of NH_3 absorbed is directly related to fineness of superphosphate. If surface area is more, then it comes with much more contact with NH_3.

If particle size of superphosphate is 30-40 mesh, then it requires 4 kg NH_3 per 20 kg P_2O_5. If, particle size of superphosphate is 40-50 mesh, then it requires 5 kg NH_3 per 20 kg P_2O_5. If, particle size of superphosphate is 50-60 mesh, then it requires 7 kg NH_3 per 20 kg P_2O_5. But, if the fertilizer superphosphate is very fine, there is chance of loss of fertilizer through leaching. Thus, the usual level of acidulation for SSP is chosen as 5-7 kg NH_3 per 20 kg P_2O_5 and for TSP, it is 2-3 kg NH_3 per 20 kg P_2O_5.

(2) **Moisture Content (6-8%):** It is observed that if moisture content is 6%, then 95% of added NH_3 is absorbed. If, the moisture content is 4%, then 90% of the added NH_3 is absorbed by supephosphate. If, moisture content is further reduced to 2%, the absorption of NH_3 with superphosphate drastically reduces to 80% only. Thus, it is usually maintained at 6-8% moisture content. If, it is too dry (<6%), there will be less absorption of NH_3 by the superphosphate and losses of NH_3 will increase.

(3) **Time of reaction (Time of injection):** Normally 3-4 minute time is allowed for ammoniation reaction so that 90% of NH_3 can reacts with SSP or TSP. If, more time is allowed to react NH_3 with SSP or TSP, more chances of formation of DCP or TCP will be there, which is not desirable. So, reaction time is very important.

(4) **Temperature:** Temperature is not a crucial factor for ammoniation. The reaction is itself an exothermic one. Thus, within working temperature is allowed. It does not have much impact.

(5) **Adequacy of contact:** There should be adequate contact of superphosphate with NH_3. If, NH_3 is injected through only one path or one point of contact, excess NH_3 will come in contact with less amount of superphosphate and possibility of producing TCP is more. In order to avoid this, NH_3 is injected through different points *i.e.* **multipoint injector** is used for ammoniation. Thus, there will be more contact or we can say that whole amount of superphosphate will come in contact with NH_3.

(6) **Level of Ammoniation:** How much ammoniation can be done?

For SSP, about 5.5 to 6.5 (5-7) kg NH_3 per 20 kg P_2O_5 and for TSP, 2.5 to 3.5 (5-7) kg NH_3 per 20 kg P_2O_5 are required for completing ammoniation. This is because, SSP is a low analysis fertilizer, that means 20 kg P_2O_5 comes from 125 kg SSP (20×100/16 = 125). In case of TSP, 20 kg P_2O_5 comes from 43.5 kg TSP (20×100/46 = 43.5). That is why, in case of TSP, the lesser amount of NH_3 is required or level of ammoniation is less compared to SSP.

Although it is a cheaper product for manufacturing a fertilizer containing both N and P, ammonium superphosphate is not produced in India, rather it is imported. This is because it require much more sophistication for manufacturing during the ammoniation process.

3.16. NP/NPK Complex Fertilizers

Complex fertilizer may be defined as any fertilizer containing two or more nutrients in which at least two major nutrients are in definite chemical combination. Complex fertilizers must satisfy some conditions. They must contain two or more major nutrients; and at least two major nutrients are in a complex form, or chemical combination is taken place between them.

It may be noted here that ammonium superphosphate is not a complex fertilizer. There is no definite chemical compound in ammonium superphosphate though they contain two nutrients *i.e.* N and P.

Mixed fertilizers are those fertilizers which are prepared by physically mixing of two or more straight fertilizers. Urea can be mixed with muriate of potash. This is simply known as a mixer or mixed fertilizer. Here, only physical mixing takes place.

There are two groups of complex fertilizers, namely (1) Ammonium phosphates and (2) Nitrophosphates.

(1) **Ammonium Phosphate:** It is very important group of complex fertilizer, because most complexes fall in this group. Basically, it involves a reaction product produced by reacting NH_3 with H_3PO_4. The most important fertilizers and their grade under ammonium phosphate are listed in Table 3.9.

Ammonium phosphate-based fertilizers were produced in USA as early as 1917. The term '**ammonium phosphate**' encompasses a wide variety of fertilizers produced by the ammoniation of phosphoric acid (H_3PO_4), often in admixture with other materials.

Table 3.9. Ammonium phosphate based fertilizers and their grades

Ammonium Phosphate	Grade
Monoammonium phosphate (MAP)	11-52-0 and 10-26-26
Diammonium phosphate (DAP)	18-46-0 and 20-48-0
Urea ammonium phosphate (UAP)	28-28-0, 20-20-0 and 20-24-0
Ammonium phosphate sulphate (APS) or Ammonium sulphate phosphate (ASP)	16-20-0 and 20-20-0
Ammonium phosphate sulphate nitrate (APSN)	20-20-0

**Grade:* Relative proportion of N, P and K in fertilizer product.

It may be borne in mind that NPK Complex is not a complex in real sense. The real complex is in NP and K is introduced to it by simple mixing of K-salt like MOP (KCl). Here, KCl remains as such.

(2) **Nitrophosphates:** In nitrophosphates compound, a considerable amount of N is in the form of nitrate which is due to acidulation of rock phosphate with HNO_3. There are very few fertilizers in this group. Two important grades of this group are 15-15-15 and 20-20-0. The other important grades are 13-13-20 and 23-23-0.

3.16.1. Advantages of complex fertilizers

(i) Complex fertilizers are called high analysis fertilizer. They have high nutrient content. Because of high analysis fertilizer, the cost of production of complex fertilizers as well as transport and applications are much less per unit of fertilizer material. Thus, more and more complex materials are producing now-a-days. More than 80% of the P comes from complex and only <20% P comes from SSP and TSP. Similarly, urea, as it is a high analysis fertilizes, 80% N comes from urea.

(ii) They have high water solubility, thus supply nutrients to plants after application.

(iii) They have good physical properties having CRH value up to 74%. There is no cake formation due to decomposition and chemical properties.

(iv) Low production cost (except raw materials), because of simple preparation method involves in mixing.

(v) They may be applied directly or combined in many ways with other materials.

3.16.2. Disadvantages of complex fertilizers

(i) In complex, N and P are supplied in a fixed proportion. But, for some crops such as legumes which need less N and more P, so it is not a suitable fertilizer to supply nutrients to legume.

(ii) Further, soil may be deficient in only one nutrient while other element is sufficient. Since complexes have fixed N and P proportion, we can not apply at an optimum amount of each nutrient through complex fertilizers.

3.17. Ammonium Phosphate

Ammonium phosphate-based fertilizers were produced in the USA as early as 1917. The term "Ammonium phosphates" encompasses a wide variety of fertilizers produced by the ammoniation of H_3PO_4, often in admixture with other materials. Such fertilizers may contain $(NH_4)_2SO_4$, NH_4NO_3 or urea as an additional source of nitrogen. The ammonium phosphate may be present as (i) MAP, (ii) DAP, or (iii) Mixture of MAP and DAP. They are used for direct application to the soil, and as an intermediate in the productions of granular mixed fertilizers and bulk blends.

In India, DAP is first produced in 1967 by Gujarat State Fertilizer Limited (GSFC). This is the grade of 18-46-0.

3.17.1. Production of Ammonium Phosphate

Orthophosphoric acid (H_3PO_4) is a polybasic or tribasic acid. Each H^+ can undergo neutralization. Thus, by neutralizing primary, secondary and tertiary position with NH_3 we can get different ammonium phosphate compounds.

$$H_3PO_4 + NH_3 \rightarrow NH_4H_2PO_4 \quad \ldots (3.50)$$

(Monoammonium phosphate, MAP) (Acidic)

$$H_3PO_4 + 2\,NH_3 \rightarrow (NH_4)_2HPO_4 \quad \ldots (3.51)$$

(Diammonium phosphate, DAP) stable (Neural)

$$H_3PO_4 + 3\,NH_3 \rightarrow (NH_4)_3PO_4 \quad \ldots (3.52)$$

(Triammonium phosphate, TAP) (Alkaline)

Three different ammonium phosphates can be obtained when the tribasic H_3PO_4 is reacted with anhydrous NH_3. The formation of each compound depends on the amount of NH_3. It is observed that 0.1 *M* solution of MAP gives pH value of 4.0, while 0.1 *M* solution of DAP gives pH value of 7.0 and that of 0.1 *M* solution of TAP gives pH value of 9.7. So, aqueous solution of MAP is acidic, while DAP is neutral and TAP is basic. Of these only MAP and DAP are produced, while TAP is not produced as a fertilizer because it is an unstable compound and decomposes due to its own alkalinity and releases part of NH_3 and converts to DAP. Other two forms *i.e.* MAP and DAP are stable and produced as P-fertilizers. They both supplies N and P in the form of NH_4 and $H_2PO_4^-$ or HPO_4^{2-}.

In India, the commercial grade fertilizer MAP produced is 11-52-0, while DAP is 18-46-0. The MAP is not marketed as such, but DAP is produced and marketed. However, MAP is used as liquid fertilizer.

3.17.2. Advantages of DAP over Other Complex Fertilizers

(i) Ammonium phosphates, particularly DAP, are the most popular P-fertilizers which are either used as such or are present in several complex fertilizers.

(ii) They have high nutrient content with good physical properties and water solubility.

(iii) Again, between the two stables MAP and DAP, DAP (18-46-0) is preferred over MAP (11-52-0). It can be seen that the total nutrients ($N+P_2O_5$) in both MAP (11+52 = 63) and DAP (18+46 = 64) are almost similar. However, DAP is preferred fertilizer over MAP or other complex fertilizers. This is because N content in DAP is relatively more (18% N) than MAP (11% N) as N requirement by most of the plants is higher, particularly at planting time such that it may be provided by DAP.

(iv) However, crystalline MAP is produced on a limited scale mainly for the preparation of clear non-pressure type liquid fertilizers.

(v) Cost of production of MAP and DAP are near about same.

(vi) DAP is marketed commercially as solid fertilizer whereas, MAP is produced as liquid fertilizer.

It is found that TSP contains 46% P_2O_5 and DAP also contains 46% P_2O_5. But DAP is more popular and preferred P-fertilizer because in addition to 46% P_2O_5, DAP contains 18% N. It is the advantage of DAP over TSP. Thus, DAP is very popular fertilizer to farmers.

Usually, two grades of DAP are produced. These are 18-46-0 and 20-48-0 grade. The specifications are as follows (Table 3.10):

Table 3.10. Specifications of two grades of diammonium phosphate (DAP)

Specifications	DAP (18-46-0 grade)	DAP (20-48-0 grade)
Moisture content (max)	1%	1%
Total N (NH_4^+-N) (min)	18%	20
Total P_2O_5 (min)	46%	48%
Water soluble P (P_2O_5) (min)	41%	43%

Here, N is present as NH_4^+-N form. Grade 18-46-0 is the most common grade in India. The grade 20-48-0 is imported one, not produced in India.

Some amounts of sesquioxide (impurities) are present in H_3PO_4 (wet-process). When wet-process H_3PO_4 is used for production of MAP or DAP, water soluble P decreases to some extent. So, depending on the presence of impurities in H_3PO_4, water solubility of P decreases. There is no DAP which is having 100% water soluble P. For production of DAP, H_3PO_4 from wet-process is preferred than thermal process. Due to its contaminant, DAP gets better quality which is not possible by very pure H_3PO_4. Moreover, thermal-grade H_3PO_4 is costly and not used in fertilizer production.

3.17.3. Commercial Production of MAP and DAP

Monoammonium phosphate (MAP) and diammonium phosphate (DAP) are produced by reacting NH_3 (gas) with H_3PO_4 (aq.) as follows:

$$H_3PO_4\,(aq) + NH_3\,(gas) \rightarrow \underset{\text{(MAP) crystal}}{NH_4\,H_2PO_4} + \Delta H \quad (\Delta H = -32.2\ \text{Kcal/mole}) \quad \dots (3.53)$$

$$H_3PO_4\,(aq) + 2\,NH_3\,(gas) \rightarrow \underset{\text{(DAP) crystal}}{(NH_4)_2HPO_4} + \Delta H \quad (\Delta H = -51.4\ \text{Kcal/mole}) \quad \dots (3.54)$$

Normally, NH_3 is obtained as by-product. One such source of NH_3 is the coal industry. While, H_3PO_4 comes from wet process (green acid).

3.17.4. Production of Diammonium Phosphate (DAP)

There are several methods to produce DAP. The basic principle is that here one mole of H_3PO_4 reacts with two moles of NH_3. The reaction can be done by single step *i.e.* all the H_3PO_4 and NH_3 are added at one time. Or the reaction is taken place in two steps (stages) so that the mole ratio of NH_3:H_3PO_4 at the final stage is maintained at 2:1.

3.17.4.1. Tennessee Valley Authority (TVA, USA) Process for Granular DAP

- In the TVA process, H_3PO_4 is neutralized in a reactor, known as **pre-neutralizer** (PN), with anhydrous NH_3 and forms a slurry, which has a mole ratio of NH_3:H_3PO_4 of about 1.35:1, a point near the maximum solubility (Fig. 3.7).
- The slurry thus formed is passed to a rotary drum which is known as **Ammoniator-Granulator** (AG). Here, excess anhydrous NH_3 is also passed. Additional NH_3 is added in the granulation step to give a mole ratio of 2:1, a point of minimum solubility. The decreased solubility from 1.35:1 to 2:1 of mole ratio assists granulation.
- Since the process is exothermic, thus, some NH_3 comes out. This excess NH_3 along with H_2O vapour then transferred to pre neutralizer through scrubber.
- After ammonitor-granulator, the material is passed to cooler or dryer. The function of cooler/dryer is to reduce the moisture content of the product. Since, there is a possibility of maximum solubility, the maximum

water or moisture content must be 18-22% so that it makes a thick slurry and the slurry must be free flowing. Here, temperature goes as high as 115 °C (boiling point) without excessive loss of NH_3. The heat of reaction raises the slurry temperature to the boiling point (115 °C) and evaporates some moisture.

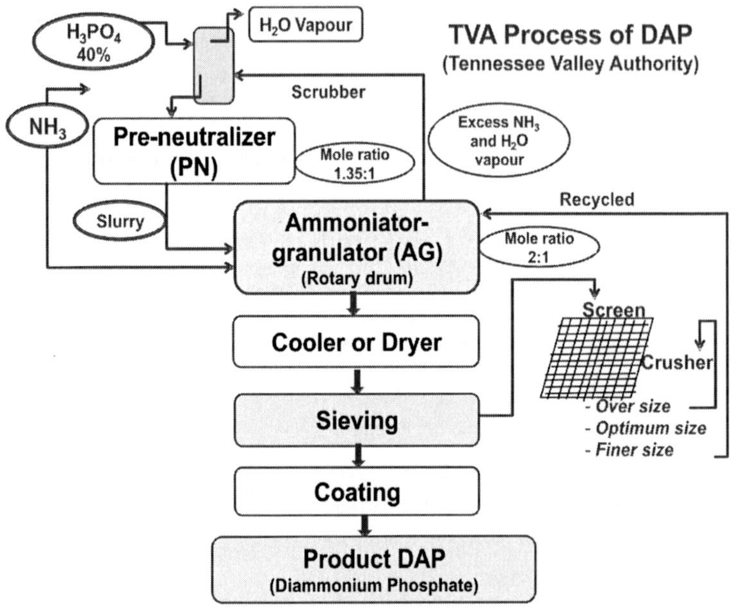

Fig. 3.7. Schematic diagram of TVA process of DAP production

- Then, the material is passed through a sieve or screen. The sieve separates the materials into 3 groups. The optimum sized material is passed to the bagging. The over sized particles are passed through a crusher and again pass to sieve. The fine sized material is again recycled to ammoniator-granulator. A slight excess NH_3 is injected in the ammoniator-granulator at this stage. The sieving is done mechanically.
- The NH_3 escaping from the granulator, dryer and pre-neutralizer is recovered by scrubbing with weak acid (30% P_2O_5) and then passes to pre-neutralizer. The usual ratio of recycle to product is 5:1. For optimum production this ratio is 2.5-4:1.

3.17.4.2. Tennessee Valley Authority (TVA, USA) Process for MAP Production

The same equipment can be used to make MAP by one of the following two procedures:

(a) In the first procedure, the pre-neutralizer is operated at a mole ratio of $NH_3:H_3PO_4$ of about 0.6:1 (a point of high solubility), and more NH_3 is added in the granulator to increase the mole ratio of 1:1 (a point of least solubility).

(b) In the second procedure, the pre-neutralizer is operated at a mole ratio of $NH_3:H_3PO_4$ of about 1.35:1; and H_3PO_4 is added in the granulator to decrease the mole ratio to 1:1.

In producing MAP, NH_3 recovered by acid scrubber is not necessary.

Advantages of TVA process are:

(i) Moisture content (18-28%) makes a thick slurry.

(ii) Material is recycled.

(iii) Granulator is common one. Here, only pre-neutralizer is added. So, by adding pre-neutralizer, we can prepare DAP.

In DAP [$(NH_4)_2HPO_4$], P is present as HPO_4^{2-} form. It is known that $H_2PO_4^-$ ion absorb better than HPO_4^{2-} ion. Still, DAP is widely used. The DAP is a popular fertilizer in India as well as other countries. It is a better P-fertilizer than SSP or TSP. As DAP contains N, the application of DAP gives green colouration of the leaves within 2-3 days. The DAP is high analysis fertilizer. For 100 kg of fertilizer materials, total nutrient content is much higher in DAP (18 kg N + 46 kg P_2O_5 = 64 kg N+P_2O_5) than straight fertilizers like ammonium sulphates (20.5 kg N), single superphosphate (16 kg P_2O_5). This gives the idea of economy of DAP in respect to storage, transport and application, which are very less in DAP. The critical relative humidity (CRH) of DAP is high (82.8) (Table 3.11). Thus, it is less hygroscopic. The physical properties are very good and better than ammonium sulphate (CRH 81) and urea (CRH 72).

3.18. Ammonium Phosphate Sulphate/Ammonium Sulphate Phosphate

A group of fertilizers known as **Ammonium phosphate sulphates (APS)** or **Ammonium sulphate phosphate (ASP)** has been popular for many years and still is going to be popular in many years. The best-known grade is 16-20-0, which is essentially consists of monoammonium phosphate (MAP) and ammonium sulphates (AS). Another grade is 20-20-0 (mixture of DAP+AS). Thus, APS/ASP is nothing but a mixture of two salts, ammonium phosphate and ammonium sulphate. It is a double salt. One reason of its popularity is that it is relatively non-hygroscopic having CRH is 76 (Table 3.11). Another advantage of APS is its S-content (12-13% S).

3.18.1. Manufacturing Process of Ammonium Phosphate Sulphates

The APS is produced by reacting H_2SO_4 and H_3PO_4 with NH_3 in a reactor (Fig. 3.8). A thick slurry is formed which is then passed to granulator. The purpose of granulator is to form the materials into granular or spherical size particles. In the normal process APS of 16-20-0 grade is produced. To produce 20-20-0 grade product, urea is passed to granulator. The granular material is then passed to cooler or dryer to reduce the moisture content to about 18-22%. The material is then passed for sieving where materials are separated into three fractions. The optimum-size is collected, bagged and marketed. The over-size fraction is passed to grinder, screened again and collected the optimum-size. The finer-size material is recycled back to granulator.

Alternatively, $(NH_4)_2SO_4$ can be used to prepare APS. Byproduct $(NH_4)_2SO_4$ is also used. For this, $(NH_4)_2SO_4$ is taken in the reactor and NH_4 is passed to it through one side. From, another side H_3PO_4 is passed to reactor. Here, a slurry is formed. Then slurry is passed to cooler/dryer. Then sieving is done.

Table 3.11. Critical relative humidity (CRH) of different fertilizers

Fertilizers	CRH* value at 30 °C
Monoammonium phosphate (MAP)	91.6 (least hygroscopic)
Diammonium phosphate (DAP)	82.8
Ammonium sulphates (AS)	81
Urea	72
Ammonium nitrate (AN)	59
Complex fertilizers	
MAP + DAP	78
Ammonium phosphate sulphate (MAP + AS)	76
Urea ammonium phosphate (MAP + Urea)	65
Ammonium phosphate nitrate (MAP + AN)	58
Urea ammonium phosphate (DAP + Urea)	62
Ammonium phosphate nitrate (DAP + AN)	59

*The values given are for pure chemical compounds; impurities present in fertilizers may lower the CRH values.

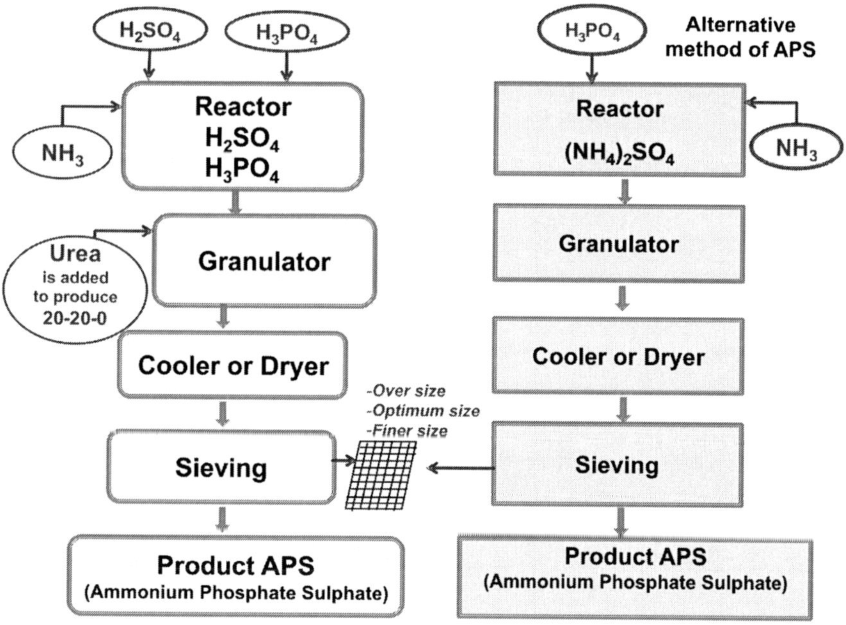

Fig. 3.8. Schematic diagram of production of ammonium phosphate sulphates (APS, grade 16-20-0)

To produce a grade of 20-20-0, usually some amount of urea is added at the granulator. Finally, the product is obtained according to the desired grade.

3.19. Urea Ammonium Phosphate (UAP) (Urea + Ammonium Phosphate)

It is the combinations of urea with ammonium phosphate. It is produced in Japan for many years. Ammonium phosphate is supplied either as a slurry produced in the plant by ammoniation of H_3PO_4 or in solid form as MAP or DAP. Urea is supplied as a solid (crushed prills or crystals), as a melt, as a concentrated solution. Ammonium phosphate is produced in the form of a slurry by neutralizing H_3PO_4 with NH_3 in the desired ratio, to which urea is added as a solid melt or concentration solution. The mixture is then granulated to UAP grades such as 28-28-0, 20-20 0 and 20 21 0. Because of presence of urea, the physical condition of UAP (CRH 65 and 62 when MAP and DAP is added, respectively) is not good as DAP. It can absorb moisture from atmosphere as it is somewhat hygroscopic.

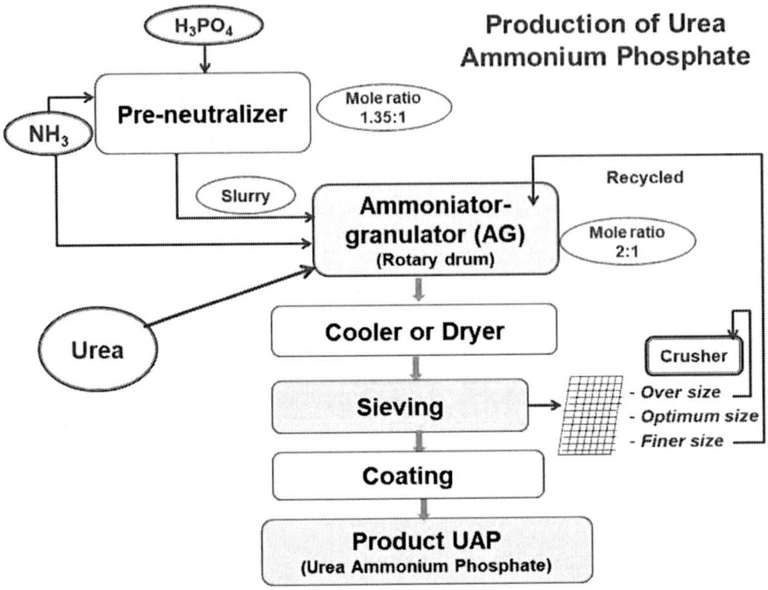

Fig. 3.9. Schematic diagram of production of urea ammonium phosphate (UAP)

The H_3PO_4 and NH_3 are allowed to pass to a pre-neutralizer as in case of DAP production (Fig. 3.9). A slurry is produced which is passed to granulator. Here, again NH_3 is passes to granulator. Thus, a solid mass is produced. This material is then passed to dryer or cooler where drying is done. Then sieving is done. Then coating is done and ultimately UAP product is obtained. Here, N is present as NH_4^+-form, but some amount is present as amide form because of addition of urea.

3.20. Ammonium Phosphate Nitrate (APN) (Grade 17-17-17)

Ammonium phosphate nitrate (APN) is a complex fertilizer containing ammonium phosphate and ammonium nitrate. The APN is produced by reacting NH_3 with H_3PO_4 and HNO_3 acids in a reactor, known as pre-neutralizer (PN). The slurry is produced by ammoniation of H_3PO_4 and HNO_3 acids. Alternatively, concentrated ammonium nitrate solution (97% NH_4NO_3) and ammonium phosphate slurry are fed to a rotary drum or granulator. Ammonium phosphate slurry is produced by ammoniating H_3PO_4 at NH_3:PO_4 mole ratio of 0.6:1 in a pre-neutralizer. More NH_3 is added in the granulator to raise the mole ratio to 1.05:1. The principal product is of 17-17-17 grade. Different grades are available across the globe. In Romania, 23-23-0 grade is produced, while TVA produces 25-25-0, 30-10-0 and 17-17-17 grades.

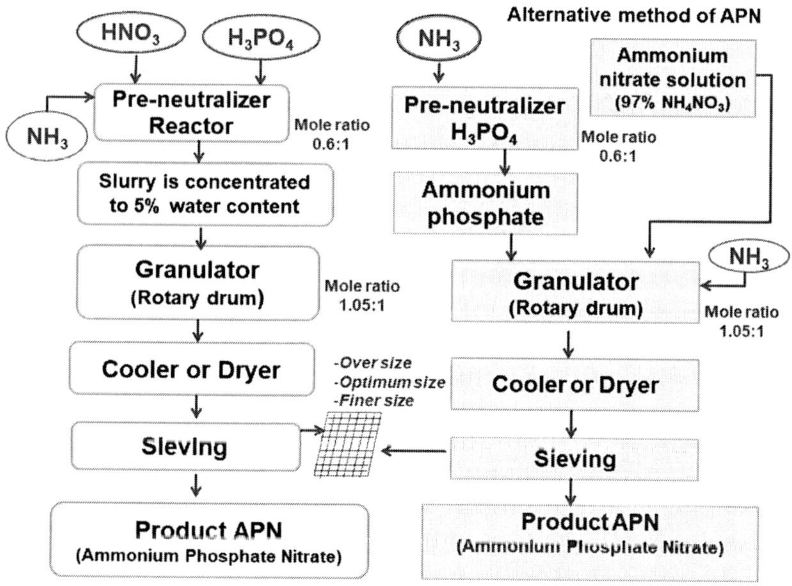

Fig. 3.10. Schematic diagram of production of Ammonium phosphate nitrate (APN)

In the TVA process, the slurry or concentrated solution of ammonium phosphate is prepared by reaction of NH_3 with HNO_3 and H_3PO_4 (Fig. 3.10). The slurry is then concentrated to 5% water content by evaporation. The material is then granulated in a granulator. Mixed H_3PO_4 and HNO_3 are neutralized with NH_3 in the first stage to a pH of 1.6, then concentrated and neutralized in the second stage to a pH of about 5.6 to prepare the hot (170 °C) concentrated solution for granulation.

3.21. Nitrophosphates

"**Nitrophosphate**" is a generally accepted term for any fertilizer that is produced by a process involving treatment of rock phosphate with HNO_3. The TVA, USA called this group of fertilizers as "**Nitric phosphate**" but "**Nitrophosphate**" seems to be more generally accepted term worldwide. It is a popular fertilizer in Europe, since 1945.

3.21.1. Advantages of Nitrophosphate

(i) The principal advantage of nitrophosphate process is that HNO_3 is used for the dual purpose of converting rock phosphate to a more soluble form and adding N through HNO_3 in the product.

(ii) Since, N is added through HNO_3, nitrophosphate processes enjoys an economic advantage over processes that uses H_2SO_4 or other non-nutrient reagents to solubilize rock phosphate.

(iii) Nitric acid comes from NH_3 synthesis. It is less costly than other acids.

(iv) Use of HNO_3 acidulation of rock phosphate reduces the use of H_2SO_4 acid for acidulation of rock phosphate, thereby, reduce the cost of production. Since sulphur (raw material) is not available in India, and imported from other countries, the fertilizers produced involving H_2SO_4 are also costly.

3.21.2. Chemistry of Nitrophosphate

The simplest process involves the reaction between HNO_3 with rock phosphate to produce monocalcium phosphate (MCP) and calcium nitrate as shown below:

$$Ca_{10}(PO_4)_6F_2 + 14\,HNO_3 \rightarrow 3\,Ca(H_2PO_4)_2 + 7\,Ca(NO_3)_2 + 2\,HF \qquad ...\ (3.55)$$

The reaction is analogous to that of SSP production, except that HNO_3 replaces H_2SO_4. The product contains 8% N and 16% P_2O_5. The presence of calcium nitrate makes this product hygroscopic, limiting thereby, its wide scale adoption. Moreover, it is a low analysis product which is a serious disadvantage. Another problem is that of a technical difficulty of loss of N as NO_2 by decomposition of HNO_3 (or calcium nitrate) due to the heat of reaction.

Most nitrophosphate processes involve dissolution of rock phosphate in HNO_3 (50-60% HNO_3) to form H_3PO_4 and calcium nitrate.

$$Ca_{10}(PO_4)_6F_2 + 20\,HNO_3 \rightarrow 6\,H_3PO_4 + 10\,Ca(NO_3)_2 + 2\,HF \qquad ...\ (3.56)$$

Nitrophosphate processes differ mainly in the subsequent treatment of H_3PO_4 and $Ca(NO_3)_2$ solution that is formed in the first step of the process. The general objective is to remove $Ca(NO_3)_2$ or at least precipitate it in an insoluble form so that it will not interfere with the desired course of subsequent ammoniation reactions.

Thus, all the development of nitrphosphate depends on the handling of $Ca(NO_3)_2$ in the process as both H_3PO_4 and $Ca(NO_3)_2$ are soluble. Otherwise, it is better to produce orthophosphoric acid by reacting of rock phosphate with HNO_3.

3.21.3. Handling of Calcium Nitrate

Calcium nitrate [$Ca(NO_3)_2$] as produced in reaction 3.56 can be handled by following four ways:

3.21.3.1. Cooling of Mother Liquor and Crystallization of $Ca(NO_3)_2$

Calcium nitrate [$Ca(NO_3)_2$] can be removed by cooling of mother liquor and crystallization of $Ca(NO_3)_2$. This process was invented by **Erling Johnson** in 1927, who named it as the **ODDA process** after the town of **ODDA** in **Norway**.

When rock phosphate is reacted with HNO_3, the heat of reaction raises the temperature of the solution to 50-60 °C. When the solution is cooled to 15-20 °C, about 60% of the $Ca(NO_3)_2$ crystalizes as calcium nitrate tetrahydrate [$Ca(NO_3)_2.4H_2O$]. The crystals are separated from the solutions by centrifugation or filtration.

The remaining solution, now containing only 40% of the original $Ca(NO_3)_2$, is ammoniated according to the following reaction:

$$6\,H_3PO_4 + 4\,Ca(NO_3)_2 + 2\,HF + 11\,NH_3 \rightarrow 3\,CaHPO_4 + 3\,NH_4H_2PO_4 + 8\,NH_4NO_3 + CaF_2 \quad ...(3.57)$$

According to the above reaction, half of the P_2O_5 would be present as monoammonium phosphate ($NH_4H_2PO_4$) which is water solúble and the other half would be present as dicalcium phosphate ($CaHPO_4$), which is citrate soluble form.

If the temperature goes even less *i.e.* at -5 °C temperature, then 85% of calcium nitrate can be removed by crystallization. Thus, almost all calcium nitrate can be removed by cooling or lowering the temperature. This is a physical process.

It is mentioned here that any P-fertilizer which is 90% water soluble is called water soluble P. All P-fertilizers produced as SSP, TSP or DAP are water soluble P. But, there is no nitrophosphate which is water soluble P. In most of the nitrophosphates, the water soluble P is 30-40%. Thus, they are not water soluble P. There is a direct relationship between calcium nitrate removed and the solubility. More removal of calcium nitrate means more water soluble P in the final product.

3.21.3.2. Removal of Calcium Nitrate by Ion-Exchange (Physicochemical Process)

Calcium nitrate can be removed by ion-exchange process. In this method, the H_3PO_4 and $Ca(NO_3)_2$ (phosphoric acid-calcium nitrate mixture) solution resulting from the reaction of rock phosphate with HNO_3, is passed through a K-loaded resin, which absorbs calcium (Ca^{2+}) and releases potassium (K^+)

ion. The resin is regenerated by a KCl solution discharging $CaCl_2$ to waste. A simplified flow diagram of the process is shown in Fig. 3.11.

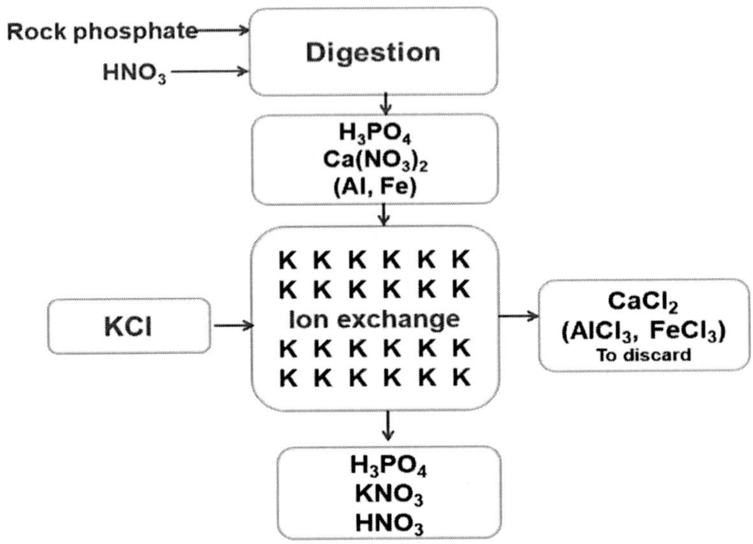

Fig. 3.11. Ion exchange column for removal of calcium nitrate

Ion exchange process is nothing but a process where an organic compound is used as resin. Here, a particular ion is exchanged by other ion. If, a column is saturated with K, then it acts as a K-column. When, the mother liquor containing H_3PO_4 and $Ca(NO_3)_2$ is passed through this K-column (resin), ion exchange occurs where Ca^{2+} replaces K^+ and KNO_3 is formed.

$$Ca(NO_3)_2 + 2\,K^+ \rightarrow Ca^{2+} + 2\,KNO_3 \quad \ldots (3.58)$$

$$\text{or, } Ca(NO_3)_2 + 2\,KCl \rightarrow CaCl_2 + 2\,KNO_3 \quad \ldots (3.59)$$

The Ca^{2+} ion is adsorbed in the column and KNO_3 comes out along with H_3PO_4. Since, there is an exchange between K^+ and Ca^{2+}, this process is called ion-exchange or more correctly cation exchange.

This KNO_3 is then separated from H_3PO_4 as KNO_3 is not as problematic as $Ca(NO_3)_2$. The column can be regenerated by passing saturated KCl solution and thus Ca^{2+} which was adsorbed on the column will come out as $CaCl_2$.

The disadvantage of this method is the high cost of resin. Thus, no single plant in the world is utilizing this method for commercial production of nitrophosphate. This process is called physiochemical process. If, the rock phosphate contains impurities (like Al and Fe) other than Ca, then Al and Fe also can be removed by ion-exchange process.

After ion-exchange, the solution which contains H_3PO_4, KNO_3, some excess HNO_3 and perhaps some residual $Ca(NO_3)_2$ is ammoniated, dried, and granulated. Thus, nitrophosphates having grade 21-9-12 and 17-17-17 are produced.

Advantages of ion exchange process are:

(i) It produces a product having low content of chloride (<0.5% Cl). Such product is desirable for crops sensitive to Cl and in saline soils.

(ii) This method has the ability to use rock phosphate of high Fe and Al-content that are eventually removed during ion exchange process.

3.21.3.3. Precipitation of Calcium as $CaSO_4$ by the Addition of Soluble Sulphate and its Removal

The solution obtained by reaction of HNO_3 with rock phosphate can be treated by the addition of soluble sulphate to precipitate part or nearly all of the calcium as $CaSO_4$ (gypsum, hemidydrate). In commercial processes $(NH_4)_2SO_4$, K_2SO_4 and H_2SO_4 have been used.

$$Ca(NO_3)_2 + MSO_4 \rightarrow CaSO_4 + 2\,MNO_3 (\text{where, } M = K^+, Na^+, NH_4^+) \quad \ldots (3.60)$$

$$\text{or, } Ca(NO_3)_2 + K_2SO_4 \rightarrow CaSO_4 + 2\,KNO_3 \quad \ldots (3.61)$$

$$\text{or, } Ca(NO_3)_2 + (NH_4)_2SO_4 \rightarrow CaSO_4 + 2\,NH_4NO_3 \quad \ldots (3.62)$$

$$\text{or, } Ca(NO_3)_2 + H_2SO_4 \rightarrow CaSO_4 + 2\,HNO_3 \quad \ldots (3.63)$$

Thus, by using some soluble sulphates, $Ca(NO_3)_2$ can be precipitated and removed. Principle involved here is a double decomposition where $CaSO_4$ is precipitated and salt of nitrate is produced.

3.21.3.4. Adjustment of Mole Ratio of $CaO:P_2O_5$

This process utilizes addition of soluble phosphate to adjust the mole ratio of $CaO:P_2O_5$ to a desired level instead of removing calcium nitrate $[Ca(NO_3)_2]$ from the system. Instead of removing Ca, the $CaO:P_2O_4$ mole ratio may be adjusted to a desired level by adding H_3PO_4 or a soluble phosphate such as ammonium phosphate. This is the basis of "phospho-nitric" or "mixed-acid" process. Thus, by adding some water soluble P, the mole ratio is adjusted and thus some amount of calcium nitrate is converted to other form of nitrate which can be subsequently separated.

$$Ca(NO_3)_2 + H_3PO_4 \rightarrow CaHPO_4 + 2\,HNO_3 \quad \ldots (3.64)$$

Technology available today, uses either (i) addition of soluble sulphate or (ii) adjusting mole ratio of $CaO:P_2O_5$ to a desired level. Still it requires some sulphur, but the extent is very less compared to SSP and TSP. So, total dependence on S is not eliminated. In ODDA process, it can be done but costly.

It may be noted here that ammonium phosphate (20-20-0) requires 1.64 tonnes of S per tonne of P_2O_5 while ammonium phosphate (25-25-0) requires 0.94 tonnes of S per tonne of P_2O_5 and urea ammonium phosphate (28-28-0) requires 0.94 tonnes of S per tonne of P_2O_5. In case of triple superphosphate, it requires 0.76 tonnes of S per tonne of P_2O_5.

3.21.4. Production Technology of Suphala (Carbonitric process) by RCFL

The Rashtriya Chemicals and Fertilizers Limited (RCFL), Mumbai followed a technology which is known as **carbonitric process** (Fig. 3.12). They modified the process themselves, but the removal of calcium nitrate does not fall in the above four categories. The rock phosphate and HNO_3 are passed to first reactor and then to second reactor or chamber where DAP and NH_3 are injected. A slurry is formed which is then passed to third chamber where again NH_3 is

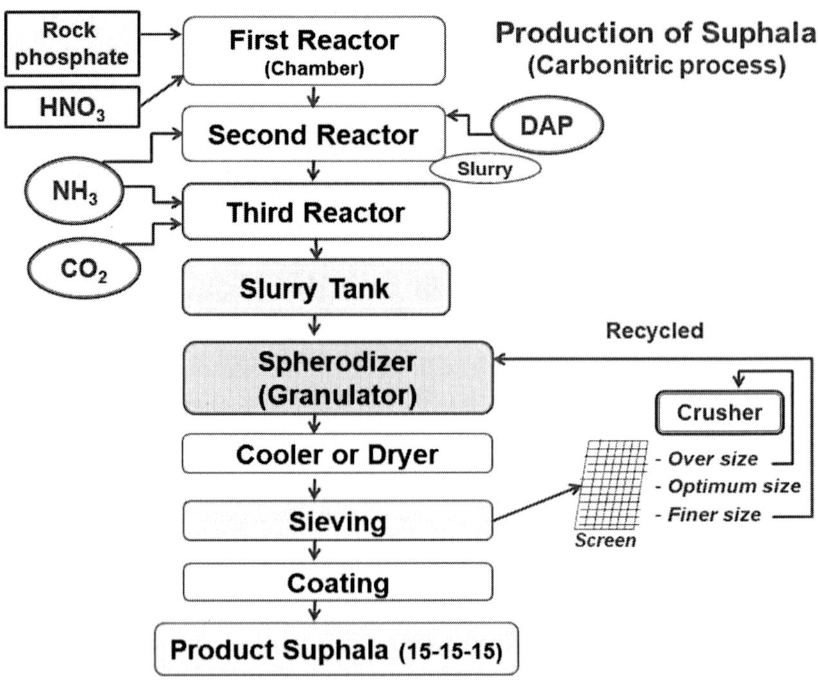

Fig. 3.12. Schematic diagram of Suphala production by RCFL

injected. Here, some muriate of potash is also introduced. Then the slurry is passed to a slurry tank. After that the product is passed through a spherodizer, where different fractions are separated into spheres, and then sieving is done, followed by cooling. After cooling, coating is given and finally product is formed. Coating is given to differentiate the two different grades 20-20-0 (if, K is not added) and 15-15-15 (if, K is added). Suphala (grade 15-15-15) is a better fertilizer or suitable for long duration crops like sugarcane, but high yielding cereals it is not suitable. Water solubility in these grades is only 30%.

Calcium nitrate separated can be used as such as fertilizer $Ca(NO_3)_2$, but often it is converted into NH_4NO_3 and $CaCO_3$ using CO_2 and NH_3. Both $Ca(NO_3)_2$ and NH_4NO_3 can be worked as N-fertilizer.

$$Ca(NO_3)_2 + 2\,NH_3 + CO_2 + H_2O \rightarrow 2\,NH_4NO_3 + CaCO_3 \quad \text{... (3.65)}$$

Carbonitric process has following four steps:

(i) Dissolving rock phosphate in HNO_3.

(ii) Ammoniating the solution from the first step to precipitate essentially all P_2O_5 as dicalcium phosphate (DCP), leaving the remainder of the calcium in solution as $Ca(NO_3)_2$.

(iii) Adding NH_3 and CO_2 to convert the $Ca(NO_3)_2$ to NH_4NO_3 and $CaCO_3$.

(iv) Granulating and drying. Final product consists of a mixture of DCP, NH_4NO_3 and $CaCO_3$ with 16-14-0 grade.

After taking calcium nitrate by any one or more of the above techniques, ammoniation is carried out under defined controlled conditions. The final slurry after ammoniation is granulated with or without addition of potassium salts by a variety of methods. Nitrophosphates are produced with different water solubility (30-70% of total P soluble in water). The grades of nitrophosphates are: 13-13-20, 15-15-15, 20-20-0 and 23-23-0.

Reactions involved are:

In the first reactor, reaction occurs as:

$$Ca_{10}(PO_4)_6F_2 + 20\,HNO_3 \rightarrow 6\,H_3PO_4 + 10\,Ca(NO_3)_2 + 2\,HF \quad \text{... (3.66)}$$

In the second reactor, reaction occurs as:

$$6\,H_3PO_4 + 10\,Ca(NO_3)_2 + 2\,HF + 7(NH_4)_2HPO_4\,(DAP)$$

$$\rightarrow 14\,NH_4NO_3 + 3\,Ca(NO_3)_2 + 6\,Ca(H_2PO_4)_2 + CaF_2 + H_3PO_4\,(slurry) \quad \text{... (3.67)}$$

In the third reactor, reaction occurs as:

$14\ NH_4NO_3 + 3\ Ca(NO_3)_2 + 6\ Ca(H_2PO_4)_2 + CaF_2 + H_3PO_4$ (slurry) $+10\ NH_3$ (anhydrous)

$$\rightarrow 20\ \underset{(AN)}{NH_4NO_3} + 4\ \underset{(MAP)}{NH_4H_2PO_4} + 9\ \underset{(DCP)}{CaHPO_4} + \underset{(Inert)}{CaF_2} \qquad \ldots (3.68)$$

The final product contains 9 moles of DCP ($CaHPO_4$) and 4 moles of MAP ($NH_4H_2PO_4$) as well as 20 moles of NH_4NO_3 and 1 mole of inert CaF_2. From the composition of mixture, it can be seen that the final product contains water soluble P (WSP) *i.e.* $NH_4H_2PO_4$, water insoluble P (WISP) but citrate soluble P (CSP) *i.e.* $CaHPO_4$ and citrate insoluble P (CISP) *i.e.* CaF_2. That is why the solubility of nitrophosphate is only 30%. Another limitation of this product is that it cannot be stored in humid condition for long time as it contains NH_4NO_3 which may deteriorate the physical conditions (hygroscopic) of the product due to formation of cakes or aggregates.

3.22. Use of Low-grade Rock Phosphate on Different Types of Soils

3.22.1. Potentiality of Rock Phosphate

Till 1960, it was believed that there was no deposit of rock phosphates in India. But in 1960, some rock phosphate deposits are found in India. Presently, the India's total reserves/resources in India are 312.67 Mt (Indian Minerals Yearbook, 2019a). However, if the Indian reserves are considered from the 'economic mineability' angle, then these will undergo further substantial reduction of 46 Mt in 2019 as compared to the world reserve of about 69 billion tonnes during the same period. The current production of rock phosphate in India is 1.25 Mt. Moreover, only a fraction of Indian rock phosphates meets the specifications of fertilizer industry. Most of the Indian rock phosphates are of low-grade, and not suitable for commercial production of P-fertilizers.

Of the total rock phosphates production across the globe, about 90 to 95% is used in fertilizer industry. Rest of the rock phosphates is used in other industries like (i) Medicine or Pharmaceuticals, (ii) Animal feed, and (iii) Water conditioners, *etc.*

3.22.2. Important Mining Agencies of Rock Phosphate in India

In India, the most important mining agencies of rock phosphates are:

(i) PPCL (Pyrites, Phosphates and Chemicals Ltd., Dehradun)

(ii) RSMML (Rajasthan State Mines and Minerals Ltd., Udaipur, Rajasthan)

(iii) MPSMC (Madhya Pradesh State Mining Corporation Ltd., Bhopal, Madhya Pradesh)

(iv) WBMDTC (West Bengal Mineral Development and Trading Corporation Ltd., Kolkata, West Bengal)

As far as Indian rock phosphates are concerned, only a small portion of indigenous reserves meets the specification of the fertilizer industry because of their low P content (low-grade) and low reactively. Another major problem in using indigenous rock phosphates is wide variation in composition among the different deposits. Even within the some deposits, they vary from lot to lot. Thus, Indian rock phosphate does not permit its use in the commercial P-fertilizers.

The feasibility of use of low-grade rock phosphates depends on their reactively that means their potentiality to release plant utilizable P in soil, which in turn depends on various factors as discussed below:

(i) **Origin of rock phosphate:** Rock phosphates occurring in deposits of igneous and metamorphic origin are less reactive, as they tend to exist in well-developed crystalline form. Consequently, results in poor performance when applied to soil as source of P. Sedimentary rock phosphates are soft materials possessing microcrystalline structure and comprise younger apatite, formed mainly in marine environment. These materials can be used for direct application in the soil.

(ii) **Mineralogy** (crystallography) of rock phosphate that means isomorphous substitution of CO_3^{2-} for PO_4^{3-} in fluorapatite and francolite as elaborated in details earlier (in section 3.5.2).

(iii) **Chemical composition of rock phosphate:** Based on P_2O_5 content, the rock phosphate reserve may be classified as high-grade (>30% P_2O_5), medium grade (25-30% P_2O_5) and low-grade (11-25% P_2O_5). The high-grade reserve can be used for wet process production of fertilizers. While, medium grade can be used for partially acidulated phosphate rock (PAPR) where minimal quantities of mineral acid is used for acidulation and low-grade materials can be used for PAPR and enriched composting or PROM (phosphate rich organic manure).

Along with P_2O_5, a substantial amount of CaO and SiO_2 are associated as accessory minerals. But contain substantial amount of CSP which determines the availability of P in rock phosphate. As rock phosphate contains no or trace amount of WSP, its direct application to crops is restricted to only in acid soils. Compared to exotic rock phosphates, Indian rock phosphates contained much lower CSP.

(iv) **Physical characteristics rock phosphate:** Physical characters, which influence the reactivity, are particle size and porosity. Particle size plays a vital role in the effectiveness of rock phosphate. This is because finer the particle size greater will be surface area and hence the total surface that will be exposed to the soil surface is large. Thus, availability of P from rock phosphate is likely to increase depending upon the fineness of particles. However, no benefits over 100 mesh particle size of rock phosphate in crop yield or P uptake when particle size increased from 60 to 200-mesh. Thus, there seems an agreement that particle size of rock phosphate should be 100-mesh size, *i.e.* 90% material passing through 100 mesh is ideal. As far as porosity is concerned, more is the micro-pores more reactive is the rock sample and more is the effectiveness or agronomic suitability.

3.22.3. Factors Affecting Availability of Phosphorus from Rock Phosphate (Dissolution of Rock Phosphate in Soil)

The feasibility of direct application of rock phosphate in soil is judged by its extent and rate of dissolution, which in turn depends on the soil properties such as: (i) proton supply or pH; (ii) calcium and phosphate concentration in soil solution; (iii) the moisture content in soil; and (iv) soil organic matter content. Dissolution of RP is also affected by the plant factors.

(i) Proton supply or pH

It is known that the feasibility or the agronomic effectiveness of a rock phosphate is higher under acid conditions. This has led to the view that soil acidity or proton (H^+) supply is the single most important factor influencing the dissolution of a rock phosphate in soil. Dissolution of rock phosphate in soil solution using fluorapatite as an example may be represented as follows (Hammond *et al.* 1986):

$$Ca_{10}(PO_4)_6F_2 + 12\,H^+ == 10\,Ca^{2+} + 6\,H_2PO_4^- + 2\,F^- \quad \ldots (3.69)$$

Thus, dissolution of rock phosphate in soil, according to law of mass action, could be favoured under conditions of low (a) soil pH, (b) soil exchangeable Ca, and (c) P concentration in soil solution.

(ii) Soil calcium and phosphate status

For increased dissolution of rock phosphate in soils, the concentration of Ca and phosphate should be minimum in the soil solution in order to drive the forward reaction. According to the solubility product principle, thus, it requires a sink for Ca and phosphate for the continued dissolution of rock phosphate. Soil acts as a sink for Ca and therefore, the affinity of a soil for Ca is high when per cent Ca is low, while the affinity is low when Ca is high in soil. Thus, the dissolution of rock phosphate is low when a soil is limed which increases soil pH and exchangeable Ca.

Dissolution of rock phosphate in soil is affected by the concentration of P in the soil solution such that high P concentration retards the process and low P concentration favours it. In fact the dissolution of rock phosphate will increase as a result of reduction of P concentration in the soil which in turn is due to (a) diffusion of P into a greater soil volume around the granule, (b) removal of P from soil solution through fixation processes like precipitation adsorption, and (c) plant uptake.

(iii) Soil moisture content

Soil moisture content is another important soil factors that affects dissolution in soils, because it is required for the transport of acidic solution to the apatite surface and the transport of Ca and P owing from the apatite surface, thereby allowing additional dissolution of rock phosphate as fertilizer may poorly active in most of the year than that of moist soils.

(iv) Soil organic matter

Soil organic matter is known to enhance the dissolution of rock phosphate which may be due to the fact that upon hydrolysis of soil organic matter, it releases some organic anions like citrate, tartrate, oxalate, *etc*. that can effectively form chelate with Ca^{2+} ion in soil solution and lower the Ca^{2+} activity in soil solution. This, in turn, provides a driving force for further dissolution of rock phosphate. The mechanisms also explain the beneficial effect on solubilization of rock phosphate when incorporated during composting of organic wastes or the beneficial effect of farmyard manure on enhancing P availability from rock phosphate. Organic matter also helps in incorporating inorganic P released from rock phosphate into the organic pool of soil P, thereby offering an additional route of releasing P from rock phosphate for eventual plant uptake through mineralization. In acid soils, use of organic manures help in increasing efficiency of rock phosphate.

The CO_2 evolved during the decomposition of organic manures resulted in the formation of weak carbonic acid which dissolves rock phosphate and rendered the availability of P to crops. Thus, an increase in the efficiency of rock phosphate is obvious. The mechanisms suggested are:

(a) Formation of phosphohumic complexes which are more easy assimilable by plants;

(b) Anion replacement of the phosphate ions by humate ions; and

(c) Coating of sesquioxide particles by humus to form a protective cover and thus reduce the phosphate fixing capacity of the soil.

(v) Sesquioxide content

Sesquioxide content of the soil determines the effectiveness of water insoluble P carriers like rock phosphate as a source of P to crops. The response of P through rock phosphate is generally higher than water soluble source like single superphosphate in the soil having higher quantity of sesquioxide because of higher fixation of P in case of later.

(vi) Dissolution of rock phosphate by plant factors

Plant influence the dissolution of rock phosphate by extracting calcium and phosphate from the soil solution through their uptake mechanism, thus promoting further dissolution through the process of acidulation by the roots. Nature of crops and species has considerable differences in their abilities to utilize P from applied rock phosphates. Among the crop species, legumes are known to be better utilizers of P from rock phosphate than non-legumes through accelerated dissolution. However, in actual practice neither all the legumes appear to feed well on rock phosphate manuring, nor all the non-legumes are unsuitable to utilize P from rock phosphate. Irrespective of the legumes and non-legumes, it is the character and extent of root system that is important in utilizing P from soil solution as they provide large contact of surface area of roots. They are also physically in close proximity to many more rock phosphate particles than those with tap root systems. The plant species having greater ability to bind Ca of rock phosphate owing to the abundance of chelating agents such as oxalates, citrate, *etc.* in their roots may be better feeders of rock phosphate. Thus, due to the greater uptake of Ca by legume roots, they are considered as better feeder. It also observed that rock phosphate is more available to dicotyledonous plants than to monocotyledonous ones and that the amount of Ca absorbed by the former was an important factor determined their ability to utilize rock phosphates.

Complete dissolution of a directly applied rock phosphate does not generally occur in a single season. Dissolution will continue beyond the year of application and therefore, application of rock phosphates, have a potential residual value. Thus, plantation crops like tea, rubber, coffee, coconut, oil palm, *etc.* because of their perennial nature, are believed to utilize P from rock phosphate more efficiently. Moreover, for them immediate P availability is not important as is for field crops. The slowly released P from rock phosphate is considered to maintain a reasonable level of available P in the soil solution.

3.22.4. Agronomic Management of Rock Phosphate

3.22.4.1. Method, Technique, Time and Dose of Application

Application technique: In contrast to water soluble P sources, which usually are band placed for localized effect near the root zone, particularly in acid soils with high P fixing capacity, rock phosphate give a better effect if applied by broadcast, and mixed well in the bulk of the soil. The reason in support of this practice is that particles of ground rock phosphate could come in intimate contact with more soil particles and thereby will be subjected to greater action of soil solvents.

Time of application: In general, the agronomic effectiveness of rock phosphate could be improved by applying it 2-3 weeks before sowing of upland crops or flooding and transplanting of rice. This is because the availability increases slowly but steadily with the passage of time of contact of phosphate particles with the soil mass, with the consequent rise in the content of available phosphate in the soil.

Rate of application: Water solubility of a phosphatic fertilizer is an important factor in deciding the rate of its application. Thus, the desirable and economical crop yields from the application of rock phosphate are only possible by applying it at least double the rate of recommended dose of phosphate for a particular crop species. It has been observed that a single application of 675 to 900 kg P_2O_5 ha^{-1} through rock phosphate once in 3-4 years to the first crop in the beginning, exhibited a clear cut edge over the more frequent and regular seasonal or annual applications in an acid Alfisols.

3.22.4.2. Possible Gains in Using Rock Phosphate as a Direct P-Fertilizer

- Besides being a rich source of P and Ca, it contains some other essential nutrients like Mg, K, S, Fe, Cu, Zn, *etc.*

- Application of rock phosphate imparts considerable liming action in the soil, because it contains about 40% CaO.
- Being an insoluble source of P, it helps in minimizing P fixation in strongly acid soils, especially, those which are rich in extractable Al and/or having higher degree of exchangeable acidity.
- Use of rock phosphate exerts considerable improvement of soil aggregation which helps in improving water retention and infiltration capacity of the soil.
- It is economical since the per unit cost of phosphate through rock phosphate is considerably lower than those of conventional phosphate fertilizers and rock phosphate could also be used successfully in neutral and alkaline soils, especially in conjunction with superphosphate or as partially acidulated materials.

3.23. Alternate Technologies for Utilization of Low-grade Rock Phosphate

3.23.1. Partial Acidulation of Low-Grade Rock Phosphate

Partial acidulation involves the use of minimal amount of mineral acid than that needed to convert rock phosphate to superphosphate (Hammond *et al.*, 1986). Low-grade indigenous rock phosphates can effectively be used by partially acidulated rock phosphate (PARP or PAPR) with nominal quantities of mineral acids compared to SSP or TSP preparation. Thus, a large portion of the acid can be saved. The reactions involving partial acidulation of rock phosphate with H_2SO_4 can be expressed as:

$$Ca_{10}(PO_4)_6F_2 + 7H_2SO_4 + 3y\,H_2O \rightarrow$$
$$3y\,Ca(H_2PO_4)_2.H_2O + 7y\,CaSO_4 + 7y\,HF + (1\text{-}y)Ca_{10}(PO_4)_6F_2 \qquad ...(3.70)$$

In this equation, the term 'y' represents the degree of acidulation, for example, 100% acidulation (production of SSP) corresponds to a 'y' value of 1.0, and a 'y' value of 0.5 refers to 50% acidulated rock phosphate.

The product of partially acidulated rock phosphate (PARP) thus contains water soluble P, *i.e.* monocalcium phosphate (MCP), citrate soluble P and citrate insoluble P. The water soluble P is considered as the starter P provided to take care of the seedling stage and the insoluble fraction of P would slowly dissolve and release P due to acidity created in the rhizosphere of the plant roots and in course of time supplies P to the crops. The extent of partial acidulation depends on the extent of soil acidity and crop to be grown. If the soil is highly acidic then 10% acidulation would be enough but when pH is higher acidulation to the extent of 20 to 50% would be desirable. Biswas and Narayanasamy (1998) reported that the effectiveness PAPRs prepared from Mussoorie, Udaipur and Purulia rock phosphates was 41-104% as effective as that of TSP in cowpea, while for wheat grown with residual fertility the effectiveness was 63-130% of TSP. Thus, such residual values had the potentiality to narrow down the performance gap between TSP and PAPR. Results also show that partial acidulation with either H_2SO_4 or H_3PO_4 to 25 to 50% was sufficient for optimum crop growth.

Mineral acids like HCl and HNO_3 in place of H_2SO_4 have been tried for production of PAPRs. In fact these could be good propositions for production of PAPR as they are the cheaper acids compared to H_2SO_4. However, the products of PAPR using HCl and HNO_3 with more than 30% acidulation are very hygroscopic and difficult to handle. The PAPRs were also made with mixtures of acids such as $HNO_3+H_2SO_4$, HNO_3+HCl and $HNO_3+H_3PO_4$ at 10 and 20% acidulation and evaluated their release in alkaline soil. All PAPRs increased the available P in soil over the rock phosphate, but did not increase the available P in soil over superphosphate.

3.23.2. Use of Rock Phosphate in Conjunction with Pyrites and/or Elemental Sulphur

The feasibility of low-grade indigenous rock phosphate could be successfully improved by using certain acidifying materials, namely elemental S, gypsum, low-grade pyrites, pressmud, *etc*. However, because elemental S is not available in India and are too expensive to be imported it is rarely used in India. The most suitable acidifying material used for improving rock phosphate is low-grade pyrites, which are available in large quantities. These are of sedimentary geological origin and pyrophoric in nature and are highly oxidizable. These have been tested and found their suitability in accentuating the efficiency of rock phosphate in alkaline calcareous soils. On addition to moist soil, pyrite gets oxidized by reacting with moisture and air to produce H_2SO_4 and ferrous sulphate as follows:

$$FeS_2 + H_2O + 3\tfrac{1}{2}O_2 \rightarrow H_2SO_4 + FeSO_4 \qquad \ldots (3.71)$$

Ferrous sulphate on hydrolysis again forms H_2SO_4.

$$FeSO_4 + H_2O \rightarrow FeO + H_2SO_4 \qquad \ldots (3.72)$$

Thus, when pyrites are mixed with rock phosphate, there is a marked acidulation effect leading to considerable solubilizing of P from rock phosphate.

$$Ca_{10}(PO_4)_6F_2 + 7\,H_2SO_4 \rightarrow 3\,Ca(H_2PO_4)_2 + 7\,CaSO_4 + 2\,HF \quad ...(3.73)$$

The first step in the oxidation process of pyrites is purely a chemical process. While, in later stage, the reaction is believed to be facilitated by several bacterial species belonging to the genus *Thiobacillus (Thiobacillus oxidans)*. To get maximum benefit from the use of pyrites, optimal and favourable conditions for the auto-oxidation and hydrolysis of pyrites are a pre-requisite. Finer particle size of pyrites, aerobic and moist conditions and sodic environment favour continuous auto-oxidation. It is also advisable to apply pyrites and rock phosphate mixture (1:1 or 1:2) in the field a few days before sowing for better results (Narayanasamy and Biswas, 1998).

3.23.3. Use of Rock Phosphate in Conjunction with Superphosphate

It is believed that 80% of the applied water soluble P gets fixed into soil within 20-25 days of application. However, rock phosphate, being an insoluble one could effectively be used as a direct P fertilizer when it is applied admixed with superphosphate (SSP or TSP) for a variety of field crops in the soil having pH range of 5.5 to 6.5. This mixture of rock phosphate and SSP/TSP could also be effectively used in neutral to alkaline soils. Mixtures are generally prepared based on the P content of the constituents. The advantage of using a mixture of RP: SSP or RP:TSP lies in the assured supply of available P at least partly from the accompanying SSP or TSP in the mixture, at the seedling and root initiation stage, when role of P supply is most critical. This early stimulation of root growth enable P from SSP or TSP accessible to the roots has been exhausted. The plants are able to sustain normal growth favoured by their developed root system which in turn, is capable of extracting P from insoluble rock phosphate on one hand and gradually but steadily solubilizing rock phosphate on the other hand.

The RP: SSP or RP: TSP mixtures have been found to give desirable results not only for the responsive crops, but also for relatively inefficient crops like wheat, maize, potato, berseem, *etc*. They also reported that RP: SSP or RP: TSP (1:1 ratio on total P basis) contains sufficient amount of water soluble P to support optimum growth in the early period of the crop which is a critical stage of the crop growth. It is suggested that the H_3PO_4 produced during the hydrolysis of MCP (present in SSP or TSP) would react more with rock phosphate and less with Al and Fe in soils. Thus, effectiveness of rock phosphate in presence of water soluble P increased by increasing the water-soluble pool of P in soil solution. Thus, the substitution of rock phosphate for SSP to the extent of 50% would result in substantial savings in fertilizer cost and improve the benefit-cost ratio of P-fertilization.

3.23.4. Mixing of Low-Grade Rock Phosphate with Water Soluble Phosphate by Compaction

Another alternative method of using low-grade rock phosphate is by the process of dry granulation (compaction) with water soluble P (WSP) fertilizers. In this technology, the steps required for granulation/compaction include introduction of feed material into the mixer followed by compaction, crushing (granulation), screening and recycling fine materials again with fresh feed into the mixer.

The compacted products are chemically equivalent to PAPR which has been found to be cheaper and more cost-effective than those produced by wet granulation process. Compaction eliminates the disadvantages associated with the application of mixtures of finely ground rock phosphate and soluble phosphates. This method, therefore, holds a lot of promise in a country like India. The compacted products of rock phosphate with SSP or with MAP + S proved to be a better source of P in different cropping sequences. Mamata *et al.* (2004) reported higher availability of P from compacted products compared to non-compacted products and more efficient in terms of apparent P recovery, P uptake, and yield increase. The WSP in compacted products can provide available P to plants at the initial stage (starter dose) that results in better plant root development which, in turn, may utilize rock phosphate more effectively at later stages than when rock phosphate is applied alone.

3.23.5. Thermally Promoted Rock Phosphate

Rock phosphates containing apatite (Ca-P minerals) are mainly used for the chemical acidulation process or direct application in soil. The non-apatite rock phosphates that contain Ca-Al-P minerals such as crandallite ($CaO.2Al_2O_3.P_2O_5.5H_2O$), wavellite [$4AlPO_4.2Al(OH)_3.9H_2O$] and millisite [$2CaO.Na_2O.6Al_2O_3.4P_2O_5.17H_2O$] due to high Al content, are not suitable for conventional chemical acidulation process. They are also not suitable for direct application due to their low reactivity. It has been reported that thermal treatment (calcination) at high temperature could increase the reactivity of such non-apatite rock phosphate due to the collapse of the crystalline structure of crandallite after calcination (Reddy *et al.*, 2000). The minerals, thus, show high solubility in neutral ammonium citrate after the water molecule of the crandallite structure is driven off and the structure becomes amorphous due to calcination at temperatures ranging from 450 to 700 °C. These aluminous phosphates were once marketed from Christmas Island (trade name "Calciphos") and Senegal (trade name "Phospal") for direct application after their thermal treatment. The results of their research showed that the calcined non-apatite rock phosphate sources can indeed be used effectively for upland and flooded rice.

3.23.6. Use of Rock Phosphate Through Composting Technology

Preparation of rock phosphate enriched compost using crop residue holds a lot of promise in developing countries like India. Preparation of rock phosphate enriched compost is based on the concept of solubilization of added rock phosphate during the process of composting. In the process of decomposition, carbonic acid and many organic acids are produced, resulting in the release of insoluble P into water and citrate soluble P which could presumably be utilized by plants. Several workers explored the possibility of increasing the availability of P from low-grade rock phosphate incorporated during decomposition of crop residues. Biswas and Narayanasamy (2002, 2006) prepared enriched compost using crop residues mixed with low-grade rock phosphate from Udaipur, Purulia, Mussoorie and Jhabua. For preparation of 1000 kg of rock phosphate enriched compost, the raw materials required are: 1000 kg of biomass (crop residues/ biodegradable wastes), 200 kg of low-grade rock phosphate (18-20% P_2O_5) and 100 kg of fresh cattle dung. The enriched compost contains about 1.4-1.5% total N and 5.0–6.0% total P_2O_5. Thus, rock phosphate-compost could be an alternative and viable technology to utilize both low-grade rock phosphates and crop residues efficiently and be used successfully as a cheaper source of P-fertilizer in place of costly water soluble P like DAP in crop production.

3.23.7. Solubilization of Rock Phosphate with Phosphate Solubilizing Microorganisms

Phosphate solubilizing microorganisms (PSM) have potential to solubilize P from rock phosphates. The effective utilization of different rock phosphates in combination with microbial culture in soils of high pH has become the subject of investigation in recent years. Several microorganisms have potential to solubilize P from rock phosphate and make it available to crops even in neutral to alkaline soils although the capacity of microorganisms to solubilize insoluble P is highly variable. These microorganisms secrete organic acids like citric, tartaric, oxalic, acetic, malic, lactic, gluconic, α-ketogluconic, *etc.*

and enzymes like acid and alkaline phosphatase, which help in bio-transformation of insoluble P to available form. The PSMs are commonly known as P-solubilizers, which may be grouped into phosphate solubilizing bacteria (PSB), phosphate solubilizing fungi (PSF) and phosphate solubilizing actinomycetes (PSA). The Division of Microbiology, IARI, New Delhi for the first time in India produced biofertilizer known as "Microphos" containing *Bacillus megatherium* var. *phosphobacterium* (Gaur and Singh, 1982). Subsequently, some other promising strains of phosphate solubilizing bacteria such as *Pseudomonas striata, Bacillus polymyxa, B. circulans, B. subtilis;* fungi like *Aspergillus awamori, A. niger, A. candidus, A. fumigatus, Penicillium digitatum, Schanniomyceas occidentalin* and actinomycetes like *Streptomyces sp.* have been isolated at IARI, New Delhi and other places and tested under field condition at different locations. The P-solubilizers are carrier based preparation containing living cells of microorganism which help in increasing crop productivity by solubilizing insoluble P and stimulating plant growth by providing hormones, vitamins and other growth factor.

3.24. Recent Advances in the Production of Phosphatic Fertilizers

It is interesting to mention that most of the widely used soluble P-fertilizers were developed in early or mid-nineties. However, suitable methods for processing a variety of low-grade rock phosphate through partial acidulation process were developed by International Fertilizer Development Centre (IFDC) in late nineties and high analysis phosphate fertilizers like ammonium polyphosphates were developed by TVA in 1972. Now-a-days, researchers are concentrating on the development of new generation fertilizers which are having higher recovery efficiency. Some of the recent findings are polymer coated WSP fertilizers, rhizosphere-controlled fertilizer (RCF fertilizer), organic complexed superphosphate, slow-release P-fertilizers with superabsorbent property and nano phosphate fertilizers.

3.24.1. Polymer-Coated Water Soluble Phosphatic Fertilizers

Reduction in the rate of conversion of WSP to water insoluble P by means of soil fixation is the main principle of the improved efficiency of coated phosphatic fertilizers. Recently, thin coating of WSP fertilizers (DAP, MAP and TSP) with water-insoluble polymers, with or without S, as slow-release P fertilizers have been commercialized by some of the fertilizer companies (*e.g.*, trade name ***'DAP-Star'*** by Hi Fert, Australia). The USA-based fertilizer company, SFP has developed and commercialized P fertilizer coated with water-soluble polymers by the trade name ***'Avail-MAP'***. Avail-MAP is a polymer (IFDC, 1986) with a very high surface charge density that can inhibit precipitation of P by acting as a platform for sequestration of P-fixing cations, such as Ca and Mg in high pH soils and Fe and Al in low pH soils. However, little information is available on the soil chemistry of such polymer-coated P fertilizers and the mechanisms of reducing P fixation by the polymer.

3.24.2. Rhizosphere-Controlled Fertilizer

The rhizosphere-controlled fertilizer is specific fertilizer having nutrient release pattern that depends on plant activity in the rhizosphere. This fertilizer is basically composed of an organo-mineral matrix of metals (Mg or Ca, Zn or Fe or other metals)-humic phosphates. The nutrient release pattern of the fertilizer is governed by the matrix. The fertilizer has two main nutrient fractions, a water-soluble fraction or starter fraction and a rhizosphere-controlled fraction insoluble in water but soluble by the action of the rhizospheric acids released by plants and microorganisms (Erro *et al.*, 2007). The starter fraction ensures the initial establishment of plants and the rhizosphere-controlled fraction meets the nutritional needs for the rest of the plant cycle. The rhizosphere-controlled fertilizer matrix is formed by the reaction of phosphoric acid (or an acid source of phosphate) with magnesium or calcium oxide (or carbonate) and iron or zinc oxide (or carbonate) in the presence of a humic acid. It is reported that the rhizosphere-controlled fertilizer is more efficient than superphosphate in both alkaline and acid soils.

3.24.3. Organic Complexed Superphosphate

In order to enhance the agronomic efficiency of applied superphosphate by reducing P-fixation in soil, a new type of superphosphate (organic complexed superphosphate) has been developed by the introduction of organic chelating agents, preferably a humic acid (HA), into the chemical reaction of sulphuric acid and rock phosphate during the production of single superphosphate (Erro *et al.*, 2011, 2012). The product thus obtained is monocalcium phosphate complexed by the chelating organic agent through Ca bridges.

Organic complexed superphosphate is produced by simultaneous addition of milled rock phosphate (80% $<100\ \mu m$), diluted H_2SO_4 (75%, w/w) and specific concentration of humic acid (dissolved in alkaline KOH solution) fraction (so that 0.1% of organic C is maintained in final complexed superphosphate) in the mixer-reactor (120-130 °C). Reaction product (liquid) was allowed to fall on conveyor belt and it becomes solid. To convert the product into powder formulation, complexed superphosphate is passed to the disintegration system. After that the product is left for 10 days at room temperature for final curing.

The agronomic performance of different types of complexed superphosphate (produced by employing diverse humic acids with different structural features) was evaluated as P fertilizers, in comparison with that of SSP, for wheat plants cultivated in pots (under glasshouse conditions) on both acidic and alkaline soils. The effect of P fertilizer application in relation to P concentration in wheat shoot revealed that complexed superphosphate performed better than SSP in diverse soils. Such result indicates that complexed superphosphate treatments are associated with the higher plant available P in soil, probably as a result of the presence of monocalcium phosphate-humic complex that prevents P from fixation in soil.

3.24.4. Superabsorbent Polymer with Slow-Release Phosphate Fertilizer

Superabsorbent polymer with slow-release phosphate fertilizer **(SAPSRPF)** is very useful in agriculture because it serves both the purposes of moisture retention and better nutrient delivery in soil. Such a polymer with slow release phosphate fertilizer has been produced by Zhan *et al*. (2004) by esterification of polyvinyl alcohol (PVA) and phosphoric acid (H_3PO_4). Phosphoric groups grafted on PVA molecule chains (through grafting polymerization), and the content of P_2O_5 in the product was 31.2%. The slow release behaviour of phosphate from SAPSRPF in water as depicted in their study claims that about 26.5% of phosphate in SAPSRPF was released out within 24 h; on the 72th h (3rd day), the total amount of phosphate released out was about 47% and on the 672th h (28th day), the total amount of phosphate released out was nearly 79%.

3.24.5. Nanophosphate Fertilizer

The growth parameters of maize as affected by application of different sized nano-particles of phosphates were studied in a solution culture experiment with maize crop grown for 30 days, excellent growth in terms of root length and volume, shoot length, dry matter yield, P content and uptake by shoot and root were observed with hydroxyapatite (HA<200 nm) followed by tricalcium phosphate (TCP<100 nm), stone RP-3 (42 nm) and HGRP3 rock phosphate (28 nm), respectively (Kundu *et al*., 2010). For further reading, please see Section 7.6.1.

Chapter 4

Potassium Fertilizer

Potassium is the third important essential nutrient element after nitrogen (N) and phosphorus (P) required by all living organisms, including plants and animals. Potassium is essential for many metabolic activities like carbohydrate metabolism, nitrogen metabolism. It is required for the activation of over 60 enzymes in plants. It controls/regulates the activities of various essential nutrients. It is vital to photosynthesis, and synthesis of carbohydrates, starch and proteins in plants. It plays vital role in controlling the opening and closing of stomata in photosynthesis. It also helps in controlling water use by plants by regulating the osmotic pressure of guard cells during transpiration. It plays a major role in transport of water and nutrients throughout the plant in xylem. Potassium improves crop yield and quality of flowers, fruits, vegetables and other field crops in terms of size, shape, colour, taste, shelf-life, and fibre quality. Potassium stimulates the growth of strong stems and provides strength to the plant against some disease by promoting thickness of the outer cell walls. It is required for translocation of assimilates. Under reduced K supply, translocation of NO_3^-, PO_4^{3-}, Ca^{2+}, Mg^{2+} and amino acids is hampered. It also imparts resistance against environmental stresses such as drought, cold and frost. Adequate K can reduce moisture loss from growing plants, thereby giving some drought resistance. It reduces lodging of crops and enhances their winter-hardiness.

It is taken up by the plants as K^+. Being a basic element, it takes part in neutralization of the acidity in plants caused by several organic acids produced. It does not form organic compounds in the plant system. It is the most mobile element in the plant. Thus, it can move through all the plant system due to osmotic pressure. When any plant sample is dipped into water, some K comes out. This shows that it is highly mobile.

Potassium is also required for animal nutrition and human beings. Potassium is essential for many metabolic functions. It maintains salt balance between cells and body fluids. Adequate K is essential for nerve function and preventing muscle cramps. It is routinely added to many animal feeds. Since K^+ is not stored in the human body, dietary replacement is required on a regular basis. Government agencies state that diets containing foods that are good sources of K and low in Na may reduce the risk of high blood pressure and stroke.

4.1. History of Potassium (*Kalium*)

Metallic potassium (K) was first discovered by **Sir Humphry Davy** from England in 1807. He made his discovery public at the **Royal Society of London** on 19th November, 1807. The word *potassium* is derived from potash, which comes from wood or plant ashes evaporated in iron pot, hence the name "**pot-ash**". It is the primary means of manufacturing the product potash before the industrial era. The elemental symbol "K" comes from Neo-Latin *Kalium* which derived from Arabic word *Al-qalya* means *alkali*. Potassium was named *Kalium* in 1809 by **Gilbert** as for sodium (Na) to "*Natrium*", which was also discovered by Sir Humphry Davy.

Potassium is a silvery-white shiny metal with a very slight purple hue to it caused by oxidation. Upon exposure to the atmosphere, it quickly forms a gray/yellow coating of oxide/peroxide having melting point 64 °C, boiling point 774 °C and density 0.86 g cm^{-3}.

Potassium is a metal that is soft enough to be cut with a knife with little force. Elemental K cannot exist in the form of atom. Potassium metal reacts rapidly with atmospheric oxygen to form flaky white potassium peroxide (K_2O) in only seconds of exposure. Also, when K comes in contact with water it reacts and produces K-oxides (K_2O). That's why, it is kept under kerosene.

In the periodic table, K is one of the alkali metals, all of which have a single valence electron in the outer electron shell which can be easily removed to create an ion with a positive charge cation (K^+) that combines with anions to form salts. Potassium in nature occurs only in ionic salts. It is found in sea water (which is 0.04% K by weight), and occurs in many minerals such as orthoclase, a common constituent of granites and other igneous rocks.

4.2. Potassium in Soils

Potassium needed for plants comes from soil. Potassium is the 7th most abundant element present on the earth crust and constitutes about 0.5-2.5%.

Total K content in soil ranges from 10^4 to 5×10^4 kg ha^{-1} in the top 15 cm soil. This indicates that a large amount of K is present in soil compared to its annual K uptake by plants. In soil, K is present in four different forms or pools such as water soluble or solution K (1-10 ppm), exchangeable K (40-600 ppm), non-exchangeable K (50-750 ppm) and structural or mineral K (5000-25000 ppm) forms. Water soluble and exchangeable pools of soil K are directly available for plant uptake. At low levels of exchangeable K in certain soil types, non-exchangeable K can also contribute significantly to the plant uptake (Memon *et al.*, 1988; Sharpley, 1989; Basak and Biswas, 2009). The exchangeable K or available K is held by negatively charged clay minerals and organic matter in soils, while non-exchangeable K consists predominantly of interlayer K in non-expanded clay minerals such as illite and lattice K in K-minerals such as K-feldspars. The bulk of total soil K is in the mineral fraction (Sparks and Huang, 1985; Sparks, 1987; Basak *et al.*, 2017, 2020a,b). All these pools of K are in dynamic equilibrium that affects the level of soil solution K at any particular time, and thus, the amount of readily available K for plants (Fig. 4.1). Levels of soil solution K are determined by the equilibria and kinetic reactions between the other forms of soil K (Sparks, 1987; Sharpley, 1989). The rate and direction of reactions between solution and exchangeable forms of K determine whether the applied K will be leached into lower horizons, taken up by plants, converted into unavailable forms, or released into available forms (Sparks, 2000). The fate of applied K in soil is also governed by clay content and clay mineralogy of soil, and the nature of crops grown. Although non-exchangeable K reserves are not always immediately available, they can contribute significantly to the labile K pool in soil. Release of non-exchangeable K to the exchangeable form occurs when levels of exchangeable and solution K are decreased by crop removal and/or leaching and perhaps by large increases

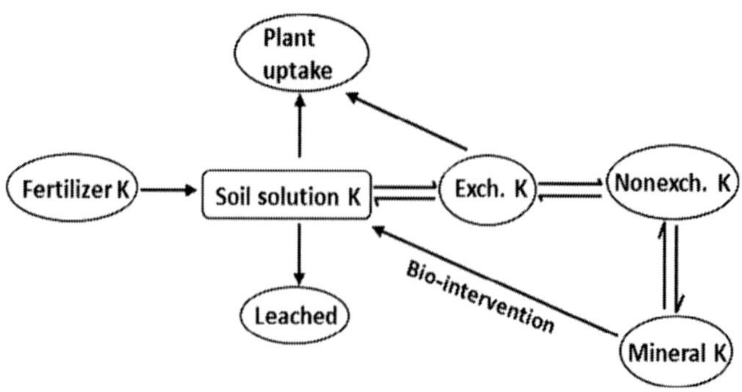

Figure 4.1. Potassium forms in soils and their transformations into soil solution through various pools and pathways

in microbial activity (Sparks, 1987). Mineral K is generally assumed to be slowly available to plants (Sparks and Huang, 1985); however, the availability is dependent on a number of factors, including the level of other forms of K, namely, solution, exchangeable and non-exchangeable, and the degree of weathering of the K-bearing minerals like feldspar and micas (Sparks, 1987).

It is evident that a static equilibrium of K never occurs because of the continuous removal of K by crop uptake and leaching. There is a continuous but slow transfer of K from the primary minerals to the exchangeable and slowly available forms. On the other hand, under some soil conditions as well as application of large amounts of fertilizer K, some reversion to the slowly available form from exchangeable and non-exchangeable forms occur. In general, readily available form accounts for 0.1 to 2%, while slowly available form accounts for 1 to 10% and total soil K 90 to 98%.

Apart from soil, a low concentration of K is found in soil water. Potassium is also found in igneous rocks. Over a long period of time, K releases from igneous rock by weathering process and deposits in the water bodies. Thus, in the coastal areas, K is accumulated. These deposits are also found in sea beds. All commercial potash deposits come from marine origin like dead seas that are now buried and salt water brines. By the process of evaporation, various salts of K are precipitated.

Potassium present in soils originates from the disintegration and decomposition of rocks containing K-bearing minerals. The K-minerals that are considered to be original sources of K are the potassium feldspars such as orthoclase and microcline ($KAlSi_3O_8$), muscovite [$KAl_3Si_3O_{10}(OH)_2$], biotite [$K(Mg,Fe)_3AlSi_3O_{10}(OH)_2$] and phlogopite [$KMg_2Al_2Si_3O_{10}(OH)_2$]. The availability of K in these minerals is of the order biotite > muscovite > potassium feldspars. Potassium is also present on soils in the form of secondary or clay minerals such as illites or hydrous mica, vermiculites, chlorites, and interstratified minerals in which two or more of these secondary clay minerals occur randomly.

The rate of K replenishment of labile pools in soil is mainly governed by the nature and weathering process of micas and feldspars. Feldspar, being a three-dimensional crystal structure with K located at interstices throughout the mineral lattice, release of K from feldspar is very difficult. It can only be liberated by destruction of the mineral. In case of micas, interlayer K can be released by exchange with other cations without alteration of the mineral.

4.3. World Reserves and Production of Potassium

In 1857, potash was first mined from waste salt deposits in Germany and since then, minerals containing K as fertilizer are being used commercially. The world reserves are estimated at approximately 3,600 Mt of K_2O content (Indian Minerals Yearbook 2019b). Reserves are located mainly in Canada (27.8%), Belarus (20.8%), Russia (16.7%), China (9.72%), USA (6.11%), Germany (4.17%) and Chile (2.78%) (Table 4.1).

The total world production of potash in 2018 was 43.2 Mt in terms of K_2O content as against 42.7 Mt in 2017 (Indian Bureau of Mines 2020). Canada is the world's leading global potash producer with annual output of 13.8 Mt. Belarus ranks the second spot in terms of total potash production at 7.35 Mt, followed by Russia (7.02 Mt), China (5.50 Mt), Germany (2.83 Mt), Israel (2.35

Table 4.1. World reserves of potash by principal countries in 2018

Country	Reserves (in Million tonnes of K_2O content)	Share (%)
World: Total (Rounded off)	>3600	
Canada	1000	27.8
Belarus	750	20.8
Russia	600	16.7
China	350	9.72
USA	220	6.11
Germany	150	4.17
Chile	100	2.78
Spain	68	1.89
Brazil	24	0.67
Israel	Large#	
Jordan	Large#	
Laos	20	0.56
Other countries	300	

Source: US Geological Survey (2020); Indian Minerals Yearbook 2019 (Part- III: Mineral Reviews), Potash (Advance Release), 58th Edition, Government of India, Ministry of Mines, Indian Bureau of Mines, Indira Bhavan, Civil Lines, Nagpur, 23, 1-5.

#Israel and Jordan recover potash from the Dead Sea, which contains nearly 2 billion tonnes of potassium chloride.

Mt) and Jordan (1.49 Mt). In terms of per cent share to world production in 2018, Canada is the leading producer of potash with 31.9% share, followed by Belarus (17.0%), Russia (16.3%), China (12.7%), Germany (6.55%), Israel (5.44%) and Jordan (3.45%) (Table 4.2). The other important K-mining countries are Chile, Spain, USA, UK and Brazil.

4.4. Commercial Potassium Fertilizers

Potash is an impure combination of potassium carbonate and potassium salts. Over 90% of World potash is used as K-fertilizer. While all the potash fertilizers contain K, there are a number of different forms in which it exists. The two most common forms are **Muriate of Potash (MOP)** and **Sulphate of**

Table 4.2. World production of potash by principal countries

Country	Production of Potash (in million tonnes of K_2O content)		Share (%) during	
	2017	2018	2017	2018
World: Total (rounded off)	42.7	43.2	-	-
Canada (Chloride)	12.6	13.8	29.5	31.9
Belarus	7.10	7.35	17.2	17.0
Russia (Chloride)	7.32	7.02	16.4	16.3
China	5.51*	5.50*	12.9	12.7
Germany (Potassic salts)	2.96	2.83	6.93	6.55
Israel (Chloride)	2.48	2.35	5.81	5.44
Jordan	1.42	1.49	3.33	3.45
Chile (Chloride)	1.20	0.99	2.81	2.29
Spain (Chloride)	0.56	0.57	1.31	1.32
USA (Potassic salts)	0.48	0.50*	1.12	1.16
UK (Polyhalite)	0.50	0.40	1.17	0.93
UK (Chloride)	0.22	0.21	0.52	0.49
Brazil (Chloride)	0.31	0.20	0.73	0.46

Source: World Mineral Production (2014-18), BGS; Indian Bureau of Mines (2020) Indian Minerals Yearbook 2019 (Part- III: Mineral Reviews), Potash (Advance Release), 58th Edition, Government of India, Ministry of Mines, Indian Bureau of Mines, Indira Bhavan, Civil Lines, Nagpur, 23, 1-5.
*Estimate

Potash (SOP). Sulphate of potash is a premium potash fertilizer free of chloride unlike MOP, the latter being harmful to plants. The SOP is used primarily on high value crops, usually leafy plants, fruits and vegetables. The MOP is commonly used on cereal crops, such as rice, wheat, *etc*. The MOP is the most concentrated form of granular K-fertilizer and typically the most cost effective. The even granule size of MOP allows for accurate spreading/application in field.

Potash is a general term that covers all the bases containing K-bearing minerals. All commercial potash deposits come originally from evaporation of deposits from ancient lake and sea beds that are often buried deep below the earth's surface in rock formations. It is important to understand that there are different types of potash deposits and that each variation of the mineral has a different use.

4.5. Raw Materials for Potassium Fertilizers Production

The major raw materials for the commercial production of K-fertilizers are the natural deposits of potassium salts (K-bearing minerals). The important K-bearing minerals are presented in Table 4.3. The principal minerals are sylvite (KCl, 63.1% K_2O); sylvinite, a mixture of sylvite (KCl) and halite (NaCl); langbeinite ($K_2SO_4.2MgSO_4$, 22.6% K_2O); kainite ($KCl.MgSO_4.3H_2O$, 18.9% K_2O); and carnallite ($KCl.MgCl_2.6H_2O$, 17% K_2O). Sylvinite, a mixture of sylvite (KCl) and halite (NaCl) crystals, is the most extensively mined mineral and is easy to process.

4.5.1. Sylvite

Sylvite (KCl) is the most common mineral used for commercial production of K-fertilizers. It is is known as the "regular potash" by the manufacturers of K-fertilizer. Sylvite is referred to the mineral name for potassium chloride (KCl). In the potash world, sylvite is also known as muriate of potash (MOP). The largest known deposits of sylvite come from Devonian evaporite basins in Saskatchewan, Canada.

4.5.2. Sylvinite

Sylvinite is a mineral found as a mixture of sylvite (KCl) and halite (NaCl). Variable amounts of sylvite and halite are present in sylvinite. It is an important source for potash. Sylvinite is the principal potash salt mixture found in the American and French deposits. It is thought to be a decomposition product of carnallite and composed largely of sylvite (KCl) and sodium chloride (NaCl), together with small amounts of magnesium chloride ($MgCl_2$) and calcium sulphate ($CaSO_4$). The potash content in sylvinite ranges from 20 to 42% K_2O.

Table 4.3. Commercially mining K-bearing minerals (ores)

Minerals	Composition	Formula	K_2O (%)
Chloride forms of K			
Sylvite	Potassium chloride (Muriate of potash)	KCl	63.1
Sylvinite	Potassium chloride, and Sodium chloride	KCl (Sylvite), $NaCl$ (Halite)	30.2
Carnallite	Potassium chloride, and Magnesium chloride	$KCl.MgCl_2.6H_2O$	17.0
Kainite	Potassium chloride, and Magnesium sulphate	$KCl.MgSO_4.3H_2O$	18.9
Hartsalz	Potassium chloride, and Magnesium sulphate	KCl (Sylvite).$MgSO_4.7H_2O$ (Kieserite)	13.3
Sulphate forms of K			
Langbeinite	Potassium sulphate, and Magnesium sulphate	$K_2SO_4.2MgSO_4$	22.6
Polyhalite	Potassium sulphate, Magnesium sulphate, and Calcium sulphate	$K_2SO_4.2MgSO_4.2CaSO_4.2H_2O$	15.5
Schoenite	Potassium sulphate, and Magnesium sulphate	$K_2SO_4 \cdot MgSO_4 \cdot 6H_2O$	23.0
Syngenite	Potassium chloride, and Sodium chloride	$CaSO_4 \cdot K_2SO_4 \cdot H_2O$	28.8
Nitrate forms of K			
Niter	Potassium nitrate	KNO_3	46.5

4.5.3. Carnallite

This form of potash is made up of potassium magnesium chloride and water (KCl, $MgCl_2.6H_2O$). Carnallite is the principal potash mineral found in Germany deposits. At present it is of secondary commercial importance because it contains only about 9-12% K_2O. As a result, it is a good type of potash to use only when there is no source of sylvite or sylvinite. Another, disadavantage of carnallite is that it is hygroscopic, thus unsuitable for long distance transportation or shipping for its use as a fertilizer. Carnallite is mined for both its potassium and magnesium content. The global distribution of this mineral is fairly vast, with occurrences in various parts in the United States. It is also found in Russia, Canada and in some regions of the Middle East.

4.5.4. Langbeinite

Langbeinite (K_2SO_4, $2MgSO_4$), like carnallite (KCl, $MgCl_2.6H_2O$), is a potassium magnesium form of potash. However, unlike carnallite, it does not contain chloride, but sulphate, giving it the name sulphate of potash magnesia (SOPM) or potassium magnesium sulphate (PMS). Usual grade of langbeinite contains 20-22% K_2O, 21-22% S and 10-11% Mg. Similar to SOP (K_2SO_4), this variation of potash is used for chloride-sensitive crops, and where there is Mg-deficiency in the soil. The K content of langbeinite is very low. Thus, it helps to keep the price of the product low. Furthermore, it is important to note that langbeinite is primarily a Mg-fertilizer. Thus, langbeinite holds a very small percentage of the potash market in the world.

4.5.5. Kainite

The mineral kainite is found in American, German and French deposits. The colour of the product may vary ranging from all shade of white, gray, yellow and red. The kainite mineral is always found in association with other minerals, and very little pure kainite ($KCl.MgSO_4.3H_2O$) occurs as such. The commercial product of kainite consists of mixture of potassium chloride, magnesium sulphate, and magnesium chloride and sodium chloride. It contains 14-22% K_2O.

4.5.6. Hartsalz

The hartsalz is not a definite mineral. It contains natural sylvite (KCl) together with kieserite ($MgSO_4.7H_2O$) and rock salt. It is found in German deposits along with kainite and sylvinite bearing mixtures. Like carnallite, hartsalz is sometimes ground and used as K-fertilizer containing 16% K_2O in Europe.

4.5.7. Polyhalite

Polyhalite deposits occur in sedimentary marine evaporates. It is a potash mineral that contains potassium sulphate. Thus, polyhalite is known for the production of sulphate of potash (SOP). Since chloride has adverse effects on the quality of foods, when it comes to fruits and vegetables, farmers prefer to use polyhalite.

Commercially produced K-fertilizers and their nutrient content are listed in Table 4.4.

Table 4.4. Commercially produced K-fertilizers and their nutrient content

Potassium fertilizer	Formula	Nutrients content
Potassium chloride or Muriate of potash (MOP)	KCl	60-63% K_2O
Potassium sulphate or Sulphate of potash (SOP)	K_2SO_4	50-53% K_2O, 17% S
Potassium magnesium sulphate	$K_2SO_4.MgSO_4$	22% K_2O, 11% Mg, 22% S
Potassium nitrate (Niter)	KNO_3	44% K_2O, 13% N
Potassium phosphates	$K_4P_2O_7$, KH_2PO_4, K_2HPO_4	22-50% K_2O, 30-60% P_2O_5
Potassium carbonate	K_2CO_3	68% K_2O
Potassium thiosulphate	$K_2S_2O_3$	25% K_2O, 17% S
Potassium polysulphide	KS_x	22% K_2O, 23% S

4.6. Mining of K-minerals

The commercial production of world's K-fertilizers comes from mining of underground K-mineral deposits followed by beneficiation. The methods employed for commercial mining of K-minerals are: (a) Shaft Mining, (b) Solution Mining, and (c) Evaporation of Natural Brine.

4.6.1. Shaft Mining

Most potash mining takes place by drilling using vertical shafts deep into the earth. The depths of the shaft vary from a few hundred meters to over 1000 meters depending on the location of the ore body. Two or more shafts are typically dug. One of the shafts is used for transporting miners and equipments, while the second shaft is used for dragging the ore to the surface. In conventional shaft mining, vertical shafts are drilled to the depth of potash deposit. Lifts are installed to provide access for equipment, workers and to remove ore.

4.6.2. Solution Mining

This method is common for K-mining in deposits of Canada. Solution mining involves injecting heated salt water through drilled holes down to the potash bed. The hot water circulates and dissolves the soluble K salts (such as sylvinite) from the ore layer. The submersible pumps lift the brine to evaporation pond where the potash crystalizes and settles to the bottom. As the liquid cools, KCl precipitates first and settles to the bottom of the tank or pond. The KCl is further processed and the NaCl remaining in solution is pumped back underground.

4.6.3. Evaporation of Natural Brines

It is used in very small number of cases. Using solar evaporation to concentrate salt from seawater is an ancient technique. However, only a few naturally occurring surface water brines (such as the Dead Sea and the Great Salt Lake) contain sufficient K to make potash extraction feasible. Specially constructed evaporation ponds are filled with brine and allowed to concentrate until salts of K, Na and Mg are precipitated.

4.7. Beneficiation and Purification of K-minerals

Most of the potash ores contain some clay and insoluble materials or impurities that require removal before further processing of the potash. Beneficiation is the process of removal of impurities such as clay and other associated minerals like halite (NaCl) present in main K-minerals (sylvite). The ore is first rinsed in a saturated salt solution that involves passing the ore through a series of highly agitated tanks to remove the minute particles of clay from the coarser ore minerals.

The beneficiation is done by following methods:

(a) Fractional crystallization

(b) Heavy media separation

(c) Froth floatation

4.7.1. Fractional crystallization

Fractional crystallization is also known as solution and recrystallization process. In this process, the ore is dissolved in hot NaCl solution (brine) for obtaining a mixture of KCl and NaCl. It may be noted that sodium chloride is unusual in that it is only slightly soluble in hot water than cold water. The beneficiation of KCl from sylvinite ore is based upon the differences in the solubilities of sodium chloride (NaCl) and potassium chloride (KCl) in hot and cold saturated brines. Thus, in solution saturated with both NaCl and KCl, the concentration of NaCl decreases as the temperature increases, while the concentration of KCl increases with increase in temperature. Thus, when temperature of the mixture of NaCl and KCl is increased, more and more of sylvinite will dissolve and the solubility of NaCl will cease after a stage. This is continued till a saturated solution of KCl is obtained. The saturated solution containing both KCl and NaCl is suddenly cooled in a vacuum and the KCl separates out as crystals while NaCl remains in the solution. Crystallized KCl can be separated out by mechanical means and it is dried, powdered and bagged as K-fertilizer.

4.7.2. Heavy media separation of KCl from NaCl

This technology utilizes the relative density of various minerals for their separation from a mixture. It is observed that when different minerals having variable densities are agitated in a solution, the heavier mineral will sink, while the higher mineral will float. Thus, in a solution with KCl having a density of 1.99 g cm^{-3} and NaCl with a density of 2.16 g cm^{-3} are agitated, KCl will float and NaCl will sink. Thus, KCl will be separated out from NaCl.

4.7.3. Froth floatation

Most important beneficiation method is froth floatation. It is a separation process in which a solid is treated with a selective agent and agitated or aerated. The treated solid adheres to bubbles of air, and the froth is floated off the suspension. Non-floating material is removed as a pulp. In case of sylvinite, the reagents commonly used for beneficiation contains a binder, alcohol and a mixture of fatty acid amines composed of mono-n-octademyl (25%), mono-n-hexadecyl (25%) and mono-n-octademyl (50%) neutralized with glacial acetic acid. The amine is used in a 1.0% water solution. The amines coat the KCl but not NaCl in solution. Air bubbles cling to amine and float KCl to surface, while NaCl and clay sink to the bottom. In order to depress the floating materials, a binder is used. The binder attracts the impurities and settles down. Alcohol acts as floating agent or frothing agents.

4.8. Dewatering and Grinding

After beneficiation, the K-mineral (sylvite) is then rinsed with saturated brine water followed by dewatering, centrifugation, drying, grinding and compaction of the finished product to desired particle sizes.

Compaction: Compaction produces granular material by compressing fine particles of hot potassium chloride in a roller press. The sheet of pressed flakes is then crushed and screened to uniform sizes.

Crystallization: The process to make pure and totally soluble potassium chloride

Hot-process: Potassium chloride is dissolved in boiling water to dissolve NaCl and KCl. As the hot brine cools, salts are differentially crystallized and removed from solution.

Cold process: The solubility of potassium chloride is lower in cold temperature than Na and Mg salts, allowing crystallization and separation.

Crushing and Grinding: Crushing is done in order to reduce the particle size to <2 mm prior to separating the potash minerals from the clay and other salts.

4.9. Potassium Fertilizer Production and Technology

As the world's population continues to grow, so too does the demand for agricultural crops. As a result, it is becoming increasingly important for farmers to grow more crops more efficiently. That's why potash fertilizers come into picture. They not only increase the quality of food being grown, but also increase crop yield. Farmers use various fertilizers to grow crops, but the two most common K-fertilizers are muriate of potash (MOP) and sulphate of potash (SOP).

4.9.1. Methods for Manufacturing of Muriate of Potash (Potassium chloride)

Potash is commercially marketed as muriate of potash (MOP) or potassium chloride (KCl). However, in potash deposits, only a small quantity is present in its mineral form (sylvite, KCl). It is obtained from underground deposits of sylvite or sylvinite after mining followed by beneficiation. The process of beneficiation is the removal of part of the impurities and upgrading the quality of the finished product. In case of sylvinite ore, usually the impurity is due to sodium chloride (NaCl) present in it. While, in case of carnallite and hartsalz, the impurities are magnesium chloride and magnesium sulphate.

The methods employed for removing impurities (other salts) from the potassium chloride consisting of treatment with hot solutions followed by cooling and crystallization. Due to their differential solubility in water at different temperatures, potassium, sodium and magnesium chlorides and sulphates dissociate themselves when the crude minerals are dissolved in water. Potassium chloride is more soluble in hot water than sodium chloride; while in cold water, potassium chloride is less soluble than sodium chloride, thus potassium chloride is precipitated and separated by filtration.

Muriate of potash is a crystalline white to pink or red or brown (due to iron ore impurities) material depending on the mining and recovery process and contains 60% K_2O, all in water soluble form giving rise to plant available K^+ ion. It is a neutral fertilizer. On application to the soil, like ammonium ions (NH_4^+), K^+ ions are also adsorbed on to the negatively charged clay particles, thus not lost through leaching. It can be applied at the time of sowing or prior to sowing as basal. On heavy soils, placement of this fertilizer in bands is better, as in the case of phosphatic fertilizers. It can be applied to all soils and crops; however some crops are sensitive to chloride ion (Cl^-) (fruit crops like citrus, grapes, almond, apricot, avocado, banana, mango, peach, berries – strawberry; vegetables like lettuce, onions, sweet pepper; field crops like potato, tobacco; and coffee and some flowers). The Cl^- ion deteriorates the quality like colour, taste and aroma which doesn't make food very appealing. If high levels of Cl^- are present in the soil or if irrigation water is being used to grow crops, adding MOP can create a toxic imbalance in plant nutrients. That means levels of the fertilizer have to be carefully managed, and use of MOP should be made with caution in those selected crops. It can be beneficial for soils that are low in Cl^-. Though it can be safely mixed with all fertilizers, but when mixed with urea it should not be stored for a long time as it may form cake which hamper uniform spreading or application. The MOP remains the most popular commercial K-fertilizers and accounts for more than 90% of K-fertilizer consumed in India for direct application and for manufacture of NPK complex fertilizers.

4.9.2. Methods of Manufacturing Sulphate of Potash (Potassium Sulphate)

Potassium sulphate (K_2SO_4) is also known as sulphate of potash (SOP). The SOP is not a naturally occurring mineral, and usually is produced through chemical methods.

4.9.2.1. Reaction of sylvite with sulphuric acid

It is prepared by treating sylvite (KCl) with sulphuric acid.

$$KCl + H_2SO_4 \rightarrow K_2SO_4 \quad ...(4.1)$$

The SOP is a white crystalline salt and contains 50% potassium as K_2O which is in water soluble form, and 18.3% S. It is considered a premium-quality K-fertilizer. It contains two key nutrients for growing crops such as K and S. It is a least hygroscopic and neutral fertilizer. It possesses good physical properties and can be safely used in the preparation of mixed fertilizers.

It can be applied to all soils and all crops at or prior to sowing. Use of SOP improves both quality and crop yields and makes plants more resilient to drought, frost, insects and even disease. The SOP has been known to improve the appearance and taste of foods and can improve a plant's ability to absorb essential nutrients like P. Most often SOP is used on high-value crops like fruits, vegetables, nuts, tea, coffee and tobacco. The SOP works better on crops that are sensitive to chloride, such as potatoes, tobacco, and also for fruit trees and vegetables. Behaviour is almost similar to KCl in soil, but has the advantages of supplying S and having a lower salt index. Because SOP is a manufactured product and also a higher-quality product, therefore, the price is more than MOP.

The most common method to produce the fertilizer SOP involves the steps of pouring potassium and other raw minerals into a muffle furnace at above 600 °C; creating a reaction between sylvite (potassium chloride, KCl) and sulphuric acid (H_2SO_4). This method accounts for roughly 50-60% of total global SOP production.

$$KCl + H_2SO_4 \rightarrow K_2SO_4 \qquad \ldots(4.2)$$

4.9.2.2. Reaction of sylvite with sulphate salts

The next most common method of producing SOP, accounting for about 25-30% of supply, is by reacting potassium chloride with various sulphate salts to form what is called a double salt. The most common raw material employed for this purpose is sodium sulphate.

$$KCl + MSO_4 \rightarrow K_2SO_4 + MCl \text{ (where, M = Mg, Na, Ca)} \qquad \ldots(4.3)$$

or, $$2\,KCl + Na_2SO_4 \rightarrow K_2SO_4 + 2\,NaCl \qquad \ldots(4.4)$$

4.9.2.3. Langbeinite Process

In this method, Langbeinite is dissolved in water to which a concentrated solution of potassium chloride is added to yield precipitates of sulphate of potash (SOP) or potassium sulphate.

$$K_2SO_4.2MgSO_4 + 4\,KCl = 3\,K_2SO_4(\downarrow) + 2\,MgCl_2 \qquad \ldots(4.5)$$

The precipitate Langbeinite of potassium sulphate (K_2SO_4) is separated from solution by decantation. The material is then dried and bagged as SOP.

4.9.2.4. Glaserite Process

Sodium sulphate is reacted with potassium chloride to yield potassium sulphate.

$$Na_2SO_4 + 2\ KCl = K_2SO_4 + 2\ NaCl \quad ...(4.6)$$

The above reaction occurs in two steps. In the first step, sodium sulphate reacts with potassium chloride to form an intermediate product known as glaserite ($Na_2SO_4.3K_2SO_4$). In the second step, glaserite is reacted with KCl to form K_2SO_4 as follows:

$$4\ Na_2SO_4 + 6\ KCl = \underset{\text{Glaserite}}{Na_2SO_4.3K_2SO_4} + 6\ NaCl \quad ...(4.7)$$

$$\underset{\text{Glaserite}}{Na_2SO_4.3K_2SO_4} + 2\ KCl = 4\ K_2SO_4 + 2\ NaCl \quad ...(4.8)$$

4.9.2.5. Burkite Process

Burkite ($Na_2CO_3.2Na_2SO_4$) is treated with sylvite (KCl) to get glaserite ($Na_2SO_4.3K_2SO_4$). Glaserite is separated and then treated with sylvite (KCl) to yield potassium sulphate, which is filtered off, dried and bagged as SOP fertilizer.

4.9.2.6. Mannheim Process

Sylvite (KCl) is treated with sulphuric acid to get potassium sulphate as follows:

$$H_2SO_4 + 2\ KCl = K_2SO_4 + 2\ HCl \quad ...(4.9)$$

4.9.2.7. Hargreaves Process

In this process, sulphur is burnt to yield sulphur dioxide.

$$S + O_2 = SO_2 \quad ...(4.10)$$

Sulphur dioxide is reacted with air, water vapour, and hot potassium chloride to yield potassium sulphate.

$$SO_2 + \tfrac{1}{2}\ O_2 + H_2O + 2\ KCl = K_2SO_4 + 2\ HCl \quad ...(4.11)$$

4.9.2.8. Phosphogypsum Process

Phosphogypsum is the byproduct of wet process of phosphoric acid. When rock phosphate is treated with sulphuric acid to produce phosphoric acid, gypsum is generated as a byproduct.

$$Ca_{10}(PO_4)_6F_2 + 10\ H_2SO_4 + 20\ H_2O = 10\ CaSO_4.2H_2O + 6\ H_3PO_4 + 2\ HF\ (\uparrow) \quad ...(4.12)$$

Phosphogypsum reacts with KCl to yield K_2SO_4 as follows:

$$CaSO_4.2H_2O + 2\ KCl = K_2SO_4 + CaCl_2 + 2\ H_2O \quad ...(4.13)$$

This reaction is carried out in ammoniacal solution, where potassium sulphate is separated out from the aqueous solution of $CaCl_2$. Conversion of phosphogypsum to K_2SO_4 is increased if KCl is used in excess of stoichiometric ratio. The solubility of K_2SO_4 is greatly reduced at higher concentration of ammonia in the reactor solution. Along with K_2SO_4, syngenite ($K_2SO_4.CaSO_4.H_2O$) is also generated. The ratio of K_2SO_4 to syngenite in the generated mass depends largely on the reaction conditions.

4.9.3. Other Potash Fertilizers

Potassium nitrate (KNO_3), potassium magnesium sulphate ($K_2SO_4.MgSO_4$), wood ash, *etc.* are other sources of potassium that can be added to the soil as K-fertilizer, but are of minor importance.

4.9.3.1. Potassium nitrate (KNO_3)

Potassium nitrate (KNO_3) is a special kind of fertilizer where N and K are supplied without the presence of chloride. It can be used as solid as well as solution types of fertilizers. However, the cost of potassium nitrate is very high. Potassium nitrate is also known as **saltpeter** or **nitre**. The alternative uses of potassium chloride other than fertilizer are pharmaceutical or numerous medical, chemical and industrial formulations.

Potassium nitrate is a white, crystalline material having 13% N and 44% K_2O. It decomposes, after melting point (308 °C) into potassium nitrite and oxygen. The physical condition is very good with high CRH value of 94. It is mostly used in fruit trees and in crops such as cotton and vegetables. It is valued when both K and N are needed for plant nutrition.

Manufacturing of potassium nitrate can be done by following three methods:

(a) Manufacturing of Potassium Nitrate from Indian Saltpetre

Crude Indian saltpetre or nitre (high in KNO_3) is dissolved in water and filtered. The filtrate is evaporated when potassium nitrate crystallizes out. It is then taken out and dried.

(b) Manufacturing of Potassium Nitrate from Chilian Saltpetre

Potassium nitrate (KNO_3) may be manufactured from Chile saltpetre (high in $NaNO_3$) and sylvite. They are dissolved in molar ratio of $NaNO_3$: KNO_3 of 17:15 in minimum quantity of boiling water.

$$NaNO_3 + KCl = KNO_3 + NaCl \quad ...(4.14)$$

The solubility of NaCl remains almost constant (35.6 g $100g^{-1}$ H_2O at 0 °C and 39.8 g $100g^{-1}$ H_2O at 70 °C); while that of KNO_3 increases with increase in temperature (13.3 g $100g^{-1}$ H_2O at 0 °C and 246 g $100g^{-1}$ H_2O at 100 °C). Water continues to evaporate from the boiling solution when NaCl separates out, which is removed subsequently. The mother liquor devoid of NaCl is suddenly cooled when KNO_3 crystallizes out.

(c) Manufacturing of Potassium Nitrate from Potassium Chloride (KCl)

Most industrially manufactured potassium nitrate is obtained by starting from potassium chloride (KCl) and nitric acid (HNO_3). Potassium chloride and nitric acid are reacted directly to yield potassium nitrate containing 30% concentration, in a reaction operating at room temperature, according to the equation:

$$KCl + HNO_3 = KNO_3 + HCl \quad \ldots(4.15)$$

In this process, an organic liquid (*e.g.* commercial grade pentanol) is used as an extracting agent for the purpose of removing HCl and shifting the reaction towards the desired direction. The dilute HCl can be used for other purposes.

4.9.3.2. Potassium Magnesium Sulphate ($K_2SO_4.MgSO_4$)

It has the advantage of supplying both Mg and S. It is frequently included in mixed fertilizers for soils deficient in Mg and S. It reacts similarly as would any other neutral salt when applied to the soil.

Potassium magnesium sulphate ($K_2SO_4.MgSO_4$) may be obtained as a byproduct of manufacturing of common salt from sea water. Here, halite crystallizes out during evaporation of sea water, and is removed. Then octadecyl-amine acetate and highly sulphonated castor oil are added to the salt solution. Then, it is thoroughly agitated in a floatation cell. The octadecyl-amine acetate and highly sulphonated castor oil form a film around each particles of the double salt $K_2SO_4.MgSO_4$ and carry them to the surface as froth, which is then skimmed off and dried.

4.9.3.3. Potassium phosphates ($K_4P_2O_7$, KH_2PO_4, K_2HPO_4)

Several potassium phosphates have been produced and marketed on a limited basis. Their advantages are that they are high analysis fertilizer having low salt index. They also contains P along with K. They are well suited for use on potatoes and other crops sensitive to excessive amounts of Cl^-; adaptable for preparation of liquid fertilizers high in K_2O.

4.9.3.4. Potassium carbonate (K_2CO_3), potassium bicarbonate ($KHCO_3$) and potassium hydroxide (KOH)

They are used primarily for the production of high-purity fertilizers for foliar application or other speciality uses. The high cost of manufacture has restricted their widespread use as commercial fertilizers.

4.9.3.5. Potassium thiosulphate ($K_2S_2O_3$) and potassium polysulphide (KS_x)

Potassium thiosulphate contains 25% K_2O and 17% S; while potassium polysulphide contains 22% K_2O and 23% S. They are compatible with most liquid fertilizers and are well suited for foliar application and drip irrigation.

4.10. Salt index

Fertilizers are categorized for their salt injury potential by the salt index, determined by placing the material in the soil and measuring the osmotic pressure of the soil solution. The salt index is the ratio of the increase in osmotic pressure produced by the fertilizer to that produced by the same weight of $NaNO_3$ having a relative value of 100. The salt index for different fertilizers is given in Table 4.5.

Table 4.5. Salt index for different fertilizers

Fertilizer	Salt index, relative to $NaNO_3$
Sodium nitrate ($NaNO_3$)	100
Potassium chloride (KCl)	116.1
Potassium nitrate (KNO_3)	69.5
Potassium sulphate (K_2SO_4)	42.6
Langbeinite ($K_2SO_4.2MgSO_4$)	43.4
Potassium thiosulphate ($K_2S_2O_3$)	68.0
Potassium hydrogen phosphate (KH_2PO_4)	8.4

4.11. Methods of Application of K-fertilizers

4.11.1. Potassium Application to Soil

Potassium fertilizers are generally applied before the crop is sown or planted. They are spread or broadcasted and mixed thoroughly with soil. However, for best management, it can be applied by band placement, particularly in high K-fixing soil. This results in reducing the area of contact between the fertilizer K and the soil. The K-fertilizers, especially muriate of potash (KCl), when come in direct contact with seeds may cause reduced germination and seedling growth. Thus, placement of K-fertilizer in band should be recommended in such a way that it is placed horizontally 5 cm away from the seed as well as about 2.5 cm below the seed to avoid salt toxicity to plant growth. Fertilizer placement above the seeds is not desirable and should not be recommended also, because then the fertilizer solution would easily move into seed zone with downward movement of water, resulting in salt toxicity. The importance of placement also decreases as higher rates of fertilizers are used. Where large quantities of K-fertilizers are to be applied, it appears best to make a part of the application as a side dressing in order to avoid injury to germination.

In some cases, it can also be applied as split application particularly in light textured soils and acid soils in high rainfall areas to reduce the losses of K by leaching. Split application is also found beneficial in highly reduced soils where conditions may hinder K uptake by crops. Usually, split application of K is followed for crops like banana, sugarcane, grapevines, papaya, pineapple and tea.

4.11.2. Potassium Application through Irrigation and Foliar Application

Most of the K-fertilizers are very water soluble and many are suitable for use in irrigation systems. Many studies have demonstrated benefits from foliar application of K to plants to alleviate K-defficiency. Foliar K applications are supplemental to the major supply of nutrients through the roots. Applications of K can alleviate mid-season deficiencies or supplement the soil supply during periods of peak demand by the plant.

4.12. Potassium Supply to the Plant Roots

Generally, potassium must be available near the root surface as well as in available form before it is absorbed into the root by its active uptake mechanisms. The mechanisms of supply of K to the root are mass flow and diffusion. Mass flow is the transport of K due to convection flow of water resulting from absorption and transpiration of water by the plant. It is normally considered to be a less important mechanism. However, mass flow could be a major mechanism for K nourishment of crops in soils naturally high in water soluble K.

Diffusion is the movement of K in response to the concentration gradient resulting in K transport from a zone of high concentration to one of lower concentration. It is a slow process in comparison to mass flow. The K diffusion to roots growing in soil occurs over short distances, usually only 1-4 mm from the root surface. Diffusion is usually the dominant and primary mechanism of K uptake by plant. The factors which affect the rate of supply of K by diffusion depends upon (i) the concentration gradient of K, (ii) the rate of diffusion, and (iii) the surface area of the root.

Potassium ion (K^+) is held around negatively charged soil colloids by electrostatic attraction as an exchangeable cation and it does not move through the soil readily with water movement in the soil. Thus, under most situations, very little K leaches from the soil. Soils in the arid regions may have higher K contents in the soil solution because of little water percolating through the profile. On the other hand, in very acid soils, presence of exchangeable iron (Fe^{2+}) and aluminium (Al^{3+}) create an unfavourable environment for the uptake of K.

4.13. Specific Considerations for use of K-fertilizers

Plant's need for K is high during the early growth stages. Hence, adequate K fertilizer should be applied before sowing. However, care must be taken to avoid salt injury to seeds and seedlings from fertilizer. For potato and tobacco, muriate of potash (MOP) should be replaced with sulphate of potash (K_2SO_4), because large amounts of Cl^- derived from MOP causes burning of tobacco leaf and decreases starch yields of potato and makes them poor for fries. As a result, crisp quality of potato chips deteriorates. Under prolonged submerged condition of soil, high dose of sulphate of potash should not be applied. Otherwise, it may induce H_2S toxicity and unavailability of cationic micronutrients, *e.g.* Fe, Mn, Zn and Cu by precipitating them as sulphides. Under unavailability of K-fertilizers, NaCl may prove effective in increasing K availability (through cation exchange of K^+ by Na^+) to K loving plants like coconut. Significant K leaching may occur in coarse textured soils, which may get accumulated in lower soil layers. Perennial crops with deep root systems are able to utilize the K accumulated in deeper soil layers. Heavy application of KCl may cause luxury consumption of K and may also cause Cl-toxicity, especially to chlorophobic crops.

4.14. Future Outlook

Agriculture is the backbone of India's economy. However, declining soil fertility impacts crop productivity. The appropriate application of fertilizer is a key factor in enhancing soil fertility and productivity and for overcoming K depletion. The market of potash is expected to increase year-on-year globally. The domestic demand is met almost entirely by imports. Thus, necessary initiatives to promote indigenous mining of potash in India must be encouraged.

The investment opportunities for potash mining in India as an import substitute mineral could be utilized for the development of mineral wealth. To carry out the feasibility and pilot plant studies for potash exploration in state of Rajasthan have been initiated by Government of Rajasthan.

Potash is an essential fertilizer nutrient that cannot be substituted. About 90% of world potash consumption is in fertilizer products. The growing world population and its need for food will require continued growth in potash production and consumption.

Potash exploration and development is expected to remain very active during the next decade. Global potash production capacity is expected to increase to 58.5 Mt yr^{-1} of K_2O in 2022 from 52.5 Mt yr^{-1} of K_2O in 2017, primarily from projects in Belarus, Canada, China, and Russia (Prud'homme, 2018). In 2017, about 25 potash-mining projects were in progress worldwide that are expected to be completed between 2017 and 2022. Other significant potash projects are under development in Australia, Brazil, Congo (Brazzaville), Eritrea, Ethiopia, Kazakhstan, Laos, Peru, Thailand, and the United Kingdom. None of these projects, however, are to be completed before 2023 (Prud'homme, 2018).

Chapter 5

Secondary Nutrient Fertilizers

5.1. Importance of Secondary Nutrients in Agriculture

Secondary nutrients such as sulphur (S), calcium (Ca) and magnesium (Mg) are the macronutrients that are required in relatively large amounts for crop growth. Among the secondary nutrients, Ca is the 5th most abundant element in the earth's crust while, Mg and S are ranked 8th and 13th, respectively. The average S content of the earth's crust ranges from 0.06 to 0.10%; whereas, Ca in soil solution of temperate region soils is 30 to 300 ppm and soils of tropical areas is 8 to 45 ppm (Tisdale *et al.*, 1997). In case of Mg, its' concentration ranges from 5-50 in temperate region and 120-2400 ppm in tropical region. The Ca and Mg are the most abundant cations occupying the exchange sites of the soil colloids. Thus, most soils with the possible exception of highly weathered, leached acid soils, contain enough Ca and Mg for crop growth. In soils, occurrence and reactions of S are different from those of Ca and Mg. The plant available S is SO_4^{2-} ion which is relatively mobile in soil solution. Like nitrogen (N), S is also subjected to chemical and biological oxidation-reduction reactions. On the other hand, Ca and Mg are much more stable and they exist in soil as Ca^{2+} and Mg^{2+} cations and present in association with soil colloidal fraction similar to that of K. In general, Ca and Mg are the most abundant cations occupying in the exchange site.

Secondary nutrients in soils may be lost due to several reasons such as (i) crop removal, (ii) leaching, (iii) erosion, and (iv) volatilization. The amount of Ca and Mg attain importance in areas where soils are either strongly alkaline or acidic. These soils are often deficient in Ca and Mg. The Ca deficiency is most widespread in acidic soils. The alluvial soils of the Indo-Gangetic Plains (IGP), being neutral to moderately alkaline in reaction, have sufficient bases on their exchange complex and Ca deficiency is, therefore, rarely found in these soils. However, alkali soils may contain inadequate quantities of Ca due to dominance of Na^+ on the exchange complex. Magnesium deficiency is found in sandy soils with low exchangeable Mg^{2+}, acid soils and those with high in native or applied K. Alkali soils have high exchangeable Na^+. The Ca^{2+} is a suitable ion to replace this Na^+ from the exchange complex when Ca^{2+} is supplied through gypsum ($CaSO_4.2H_2O$). The replaced Na^+ forms Na_2SO_4 which is then leached down to remove the salt away from the root zone of the crops. Gypsum also reacts with Na_2CO_3 to form $CaCO_3$ and Na_2SO_4 which is leached down the profile. A thorough leaching of the soil with irrigation water should be done after application of gypsum to free it from Na_2SO_4.

It has been found that Ca as a plant nutrient is more important in Ca-deficient acid soils than that as an amendment. Application of carbonate (CO_3^{2-}) and sulphate (SO_4^{2-}) salts of Ca @ 2 to 4 q ha^{-1} in furrows like a fertilizer nutrient at the time of sowing could increase the yield by about 48% (Tandon, 2012). The improvement in crop yield in sodic soil due to gypsum is because of increase is exchangeable Ca^{2+} status which indicates that Ca nutrition matter more than the decrease in soil pH. Acid soils under high rainfall areas usually show excellent crop response to Mg application. In red and laterite soils, 20% increase in yield of rice is observed with Mg application.

Sulphur and Mg are needed by plants in almost same quantities as that of phosphorus (P). The Ca requirement is even greater than that of P for many crops. The amount of secondary nutrients removed by crops depends on soil types, crop species, varieties, fertilizer sources and yield target. Generally, legumes and root crops remove more Ca than do cereals and grasses. Cereals remove about 10-20 kg Ca ha^{-1}, while *Brassica* sp. removes 150 kg Ca ha^{-1} (Tandon, 2012).

5.2. Sulphur in Crops

In India, S-deficiencies are found in almost 40% of the cultivated area (Sureshkumar and Sandeep, 2015). Sulphur is absorbed as SO_4^{2-} ion which is reduced in plants and incorporated in organic compounds. Different crop species have different S requirements. *Brassica* sp. like rapeseed mustard, cabbage, *etc.* have highest S requirement followed by legumes such as berseem, groundnut and soybean, while cereals have the least requirement of S. Oilseed crops contain appreciable S, as oil in these crops is rich in S-bearing compounds like S-glucosides. *Brassica* seed contains 1.2% S in grain and 0.13% S in straw, while legumes contain 0.25% S in grain and 0.2% in straw. Sulphur is a constituent of protein. For every 15 parts of N in protein, 1 part of S (N:S ratio of 15:1) is needed. Sulphur deficiency is mainly confirmed to oilseeds and legumes and rarely occur in grasses and cereals. In trees, S deficiency is very rare (orchard) because they take S from deeper layer. As a thumb rule, S removal per tonne of grain production is about 3-4 kg for cereals, 8 kg for pulses and 12 kg for oilseed crops. Research works revealed that application of 20-40 kg S ha^{-1} gave economical yield of crop, particularly in oilseeds and pulses (Pasricha and Sarkar, 2009). The normal uptake of S is 10-30 kg ha^{-1} depending upon crop species Sulphur uptake in cereals and grasses is 10 kg ha^{-1}, while in case of legumes, S uptake is 30 kg ha^{-1} and oilseeds is >30 kg (40-54 kg ha^{-1}).

5.3. Functions of Sulphur

(i) Sulphur is known as the 4th important essential plant nutrient after N, P and K.

(ii) Usually it is required by crops in amounts comparable to that of P.

(iii) Sulphur is a constituent of amino acids such as cysteine, cystine and methionine, hence essential for protein production.

(iv) It is involved in chlorophyll formation and activation of enzymes.

(v) It is a part of vitamins such as biotin and thiamine (B_1) and is needed for the formation of oil.

(vi) It is an essential constituent of certain volatile compound such as sulphydryl (SH) compound present in mustard oil. The pungency of mustard oil is due to this SH-compound.

(vii) It helps in oxidation-reduction system.

(viii) It increases root growth and promotes nodule formation on roots of legumes.

5.4. Deficiency Symptoms of Sulphur

Sulphur deficiency symptoms resemble to those of N deficiency. With N deficiency, the plant may become a uniform yellowish-green colour and lower leaves may turn yellow and fall from the plant (NO_3^- is more mobile). With S deficiency, the older leaves remain green and the younger leaves become yellowish-green and in severe cases become chlorotic.

5.5. Causes or Conditions of S-deficiency

(i) Intensive agricultural practices with concomitant crop removal resulted in deficiency of S in many soils.

(ii) Use of S-free high analysis NPK fertilizers and use of limited and no use of organic manures.

(iii) Complete crop removal (grain + straw/stover) from soil.

(iv) Coarse textured sandy soils generally have low total S content as compared to fine textured soils.

(v) The areas which are away from industrial activities show more S deficiency.

(vi) Deeply weathered soils which are subjected to strong leaching with high rainfall is SO_4^{2-} deficient.

(vii) In arid regions, where evaporation is more than precipitation, the chance of leaching is very less. Thus, S deficiency is very rare.

5.6. Sources of Sulphur in Soils

(i) The major sources of S under natural condition are the organic matter. More than 96% of the total S in soil is present in the organic matter under temperate conditions. However, total S in the organic form is relatively lower in tropical and sub-tropical soils. Inorganic sulphates (SO_4^{2-}) are also present in tropical and sub-tropical condition in appreciable amount. It becomes available because of decomposition of the organic matter.

(ii) Inorganic minerals present in the soils are natural minerals like sulphides of sedimentary and igneous rock. The dominant S-minerals in soils are pyrites (FeS_2, 22-24% S), arsenopyrite ($FeS_2.FeAs_2$), sphalerite (ZnS), gypsum ($CaSO_4.2H_2O$, 13-18% S), chalocopyrite ($CuFeS_2$), epsomite ($MgSO_4.7H_2O$, 12% S, 16% MgO) and galena (PbS). In acid sulphate soil of Kerala, a S-containing mineral known as Jarosite [$KFe_3(OH)_6(SO_4)_2$] and Alunite [$KAl_3(OH)_6(SO_4)_2$] are found which are formed due to aeration.

(iii) Sulphur is also present in air (atmosphere) as sulphur dioxide (SO_2). Owing to industrial activity, through combustion of fossil-fuel, SO_2 is produced and brought down during rains and made available to plants.

(iv) Sulphur (SO_4^{2-}) is also present in irrigation water.

(v) Sulphur also occurs in sea water.

Under tropical and sub-tropical conditions, organic matter content of soil is generally less than that in temperate regions. This is because whatever organic matter is added as crop stubbles or natural leaf-fall is immediately decomposed and its S is mineralized to SO_4^{2-}. The oxidation of S present in organic matter in the form of proteins and amino acids is mediated by number of microorganisms (*e.g. Thiobacillus* bacteria). Thus, in arid and semi-arid regions, much of S in these soils are inorganic form. Since 96% of the S present in organic form, the soils rich in organic matter will have high levels of S. In most of the cultivated soils, S ranges from 30 to 50 mg kg^{-1}. Among Indian soils, only 30% of total S in soil may be in the organic combination in alluvial soil (Entisols) but in Mollisol of Terai regions, it may be 70%.

5.7. Mineralization of S-compounds in Soil

The S in soil is being cycled continuously between organic and inorganic forms. The nature of compounds formed and their transformation are influenced by the biologically mediated processes. The organic S is converted to inorganic SO_4^{2-} by microbial oxidation and the process is known as mineralization.

$$\text{Organic S} \xrightarrow[\textit{(Thiobacillus)}]{\text{Oxidation } (O_2)} SO_4^{2-} + H^+ \quad \ldots(5.1)$$

The mineralization form of S can be taken up by plants and microorganisms. The rate of mineralization is affected by factors such as (i) moisture, (ii) aeration, (iii) temperature, (iv) pH, and (v) C:S ratio. The mineralization process cannot be taken place under submerged soil conditions. Thus, moisture level of <60% of field capacity is good for oxidation. The conversion of organic S to SO_4^{2-} is an oxidation process and cannot be taken place in absence of O_2. Thus, proper aeration is essential for oxidation. Under reduced conditions (*e.g.* under puddled condition in rice), SO_4^{2-} ions are reduced by anaerobic bacteria *Desulphovibrio* and *Desulphotomaculum* to sulphite and sulphide, respectively. Mineralization of S is similar to that of N. Since mineralization process is a microbial mediated process, extreme of temperature hamper this process. Thus, the optimum temperature is 35-40 °C. The formation of end product SO_4^{2-} is associated with the formation of H^+. Hence, this process lowers the soil pH. Addition of organic manures low in S content leads to proliferation of microbial growth, resulting in immobilization (*i.e.* conversion of inorganic-S to organic-S), and thereby S becomes temporarily unavailable to plants. Upon the death of microbial population, the microbial S again mineralized and becomes available (SO_4^{2-}) form. Sulphur in the soil is associated with organic carbon in a C:S ratio of about 140:1. Thus, favourable C:S ratio (about 125 to 140:1) is important for mineralization of organic matter in soils and maintenance of S availability to plants.

5.8. Functions of Calcium

(i) It is a part of the structure of cell walls and membranes and increases stiffness of straw.

(ii) It promotes early root development and growth.

(iii) It is involved in cell-division, growth and root elongation, pod filling and activation or inhibition of enzymes.

(iv) It provides a base for neutralization of organic acids, commonly termed as poisons, produced in the plants.

(v) It influences the water economy of the plant and protein-carbohydrate ratio in fat metabolism.

(vi) Effects of Ca, however, are antagonistic to those of K.

(vii) It improves intake of other plant nutrients, especially N and trace elements such as B, Zn, Cu and Mn by correcting soil pH.

(viii) Calcium encourages seed production.

(ix) For root tips Ca is more important. Calcium is the major cation in the middle lamella of cell walls of which Ca-pectate is the principal constituent. Thus, Ca provides mechanical strength to tissues.

5.9. Deficiency Symptom of Calcium

(i) In most plants root growth is reduced, root rotting frequently occurs and the roots are affected before the tops show any symptoms.

(ii) In moderate stages of deficiency, the young leaves become distorted, show spotting or neurotic areas.

(iii) In the acute stage, there is a general collapse of growing points, stunted growth and a die back of the terminal buds.

5.10. Functions of Magnesium

(i) Magnesium is an essential constituent of chlorophyll and thus vital for photosynthesis and thus essential for all green plants. It helps in maintaining the dark green colour in leaves.

$$6\,CO_2 + 6\,H_2O \xrightarrow[\text{Phosphate energy}]{\text{Chlorophyll, sunlight}} C_6H_{12}O_6 + 6\,O_2 \qquad \ldots(5.2)$$

(ii) It plays a part in the production of carbohydrate, proteins, fats and vitamins and certain catalytic reactions in the enzymatic systems.

(iii) Acts as a carrier of phosphate in plants.

(iv) Helps the translocation of starches.

(v) Regulates the uptake of other nutrients.

5.11. Deficiency Symptom of Magnesium

(i) Deficiency symptoms are frequently found on lower leaves.

(ii) In cotton, a reddening of the lower leaves occurs; while in potatoes, an orange-yellow colouration appears on the lower leaves and along the veins. In case of maize, striping or chlorosis between veins.

5.12. Materials Supplying Secondary Nutrients

There are numerous fertilizers containing secondary nutrients. In fact, fertilizer manufacturers from their very beginning manufacturing N, P and K fertilizers containing Ca, Mg and S. For example, single superphosphate (SSP) contains both Ca and S besides major nutrient P. Similarly, ammonium sulphate [$(NH_4)_2SO_4$] contains S besides major nutrient N and sulphate of potassium (K_2SO_4) contains S along with K. Calcium ammonium nitrate (CAN) supply substantial amounts of Ca beside the major nutrient N. All these fertilizers contain ample amount of secondary nutrients and discussed in Chapters 2, 3 and 4, dealing with nitrogen, phosphorus, and potassium, respectively.

Among the major plant nutrients, research work on Ca and Mg has been limited in India. However, the field management of Ca and Mg as essential plant nutrients is of considerable importance in view of the large area under acid soils where inadequate availability of Ca and Mg can be limiting factor in balanced fertilizations programme. Calcium and Mg deficiency occurs only in highly weathered acid soils. On the other hand, S has been studied extensively than Ca and Mg. Calcium and Mg deficiency is often observed in acid soils, so lime is added in those soils. Thus, Ca and Mg are automatically taken care of when we increase pH by lime. Liming is done mainly for increasing pH. There are many research works on liming but not on Ca and Mg as such. Liming is chief way to correct Ca deficiency in acid soils. Under highly sodic situation, the normal practice is to amend the soil by gypsum ($CaSO_4.2H_2O$). Lime and gypsum are not fertilizers but amendments. Soil Ca and Mg other than those added through fertilizers or liming materials come from the weathering of rocks containing different materials.

Sources of Ca and Mg are applied commonly in amounts required to adjust the pH, rather than in response to an indicated plant deficiency. The usual materials are finely ground agricultural limestone (principally $CaCO_3$) and dolomite ($CaCO_3$ and $MgCO_3$), a limestone composed principally of a double carbonate of Ca and Mg. The fineness of limestones for direct application (called liming) varies from 8-100 mesh size. The finer the materials, more is the effectiveness. Fertilizer manufacturers generally prefer a limestone with particle size of 80-100 mesh.

The benefits of liming are:

(i) ***Improved P availability***: With low pH soils and soils high in Al and Fe oxide content, P are made less available because of their reaction with these compounds. Liming to a pH of 6-7 increases P availability because of the precipitation of Fe and Al in forms other than as the phosphates. At high pH, the availability of P decreased because of formation of Ca-phosphates.

(ii) ***Nitrifications***: Most of the organisms responsible for convertion of NH_3 to nitrates require Ca.

(iii) ***Nitrogen fixation***: This process is favoured by the addition of liming agents.

(iv) ***Micronutrient availability***: Except for molybdenum (Mo), the availability of micronutrients increases with a decrease in soil pH. This can be detrimental because of the toxic nature of these elements in other than small concentrations. Depending upon the soil, a pH of 6-7 is usually most satisfactory from the standpoints of nutrient availability and toxicity.

(v) ***Physical Condition of Soil***: Liming improves the structure of fine-textured soils.

5.13. Fertilizers Containing Secondary Nutrients

Over 60 S-containing fertilizers are known to agriculture. The use of sulphur compounds to reclaim alkali soils as an amendment is commonly practiced. Finely ground calcium sulphate dihydrate ($CaSO_4.2H_2O$), known as gypsum, is used for reclamation purpose. Similarly, elemental S is also used. The details of secondary nutrients materials used as fertilizers or amendments are listed in Table 5.1.

Table 5.1. Some materials containing secondary nutrients

Nutrient Materials	Formula	Nutrient content (%)
Single superphosphate	$3Ca(H_2PO_4)_2.H_2O$ + $7CaSO_4.2H_2O$	16% P_2O_5, 9-12% S, 20% Ca, 0.3% Mg
Single superphosphate (Boronated)	$3Ca(H_2PO_4)_2.H_2O$ + $7CaSO_4.2H_2O$ + B	16% P_2O_5, 11% S, 20% Ca, 0.18% B
Triple superphosphate	$Ca(H_2PO_4)_2.H_2O$	46% P_2O_5, 14.3% Ca
Ammonium sulphate	$(NH_4)_2SO_4$	20.5% N, 24% S
Potassium sulphate	K_2SO_4	50% K_2O, 18% S, 0.6% Mg
Potassium magnesium sulphate	$K_2SO_4.MgSO_4$	22% K_2O, 20-22% S, 7-11% Mg
Potassium schoenite	$K_2Mg(SO_4)_2.6H_2O$	23% K_2O, 17% S, 10% MgO
Magnesium sulphate (Epsomite salt)	$MgSO_4.7H_2O$	12% S, 10% Mg
Zinc sulphate	$ZnSO_4.7H_2O$	11-18% S, 21% Zn
Ferrous sulphate	$FeSO_4.7H_2O$	10.5% S, 19% Fe
Manganese sulphate	$MnSO_4$	17% S, 30.5% Mn
Copper sulphate	$CuSO_4.5H_2O$	12% S, 24% Cu
Ammonium phosphate sulphate	$NH_4H_2PO_4+(NH_4)_2SO_4$	16/20% N, 20% P_2O_5, 13-15% S
Gypsum	$CaSO_4.2H_2O$	13-18% S, 16-19% Ca, 4-10% Mg
Phosphogypsum	$CaSO_4.2H_2O$	16% S, 21-29% Ca
Elemental sulphur	S	80-100% S
Sulphur-bentonite	90% S +10% bentonite	90% S
Ammonium thiosulphate	$(NH_4)_2S_2O_3$	12% N, 26% S
Iron pyrites	FeS_2	22-24% S, 19% Fe
Agricultural limestone	80-95% $CaCO_3$	32.3% Ca
Dolomite	$CaCO_3+MgCO_3$	14-32% Ca, 3-12% Mg
Calcium ammonium nitrate	$NH_4NO_3+CaCO_3$	25% N, 7-14% Ca, 4.5% Mg
Calcium nitrate	$Ca(NO_3)_2$	15.5% N, 19% Ca
Rock phosphate	$Ca_{10}(PO_4)_6F_2$	16-34% P_2O_5, 8 34% Ca
Bone meal	$Ca_3(PO_4)_2$	23% Ca, 20% P_2O_5
Basic slag	$3CaO.SiO_2$	1-2% S, 34% Ca, 3-5% Mg
Magnesite	$MgCO_3$	24% Mg
Chelated magnesium	Chelated-Mg	2-10% Mg
Magnesium chloride	$MgCl_2.10H_2O$	8-9% Mg

5.13.1. Ammonium Sulphate [$(NH_4)_2SO_4$)]

It contains 21% N and 23-24% S. Both of its N and S are in readily soluble or plant available forms as NH_4^+ and SO_4^{2-}. It is an acid forming fertilizer and its continuous use will lower soil pH and requires lime application to neutralize the excess acidity. This is one of the oldest sources of ammoniacal N, having been manufactured for many years as a byproduct of the coke or coal used in the steel industry. Approximately, one-fourth of its world production of ammonium sulphate comes from caprolactum, an intermediate raw material for the production of synthetic fibers. The acid forming reaction of $(NH_4)_2SO_4$ in soil can be advantageous in high pH-soils and for acid requiring crops such as tea. It is very good for potato to prevent scabies disease. For further details, please see Section 2.10.

5.13.2. Calcium Ammonium Nitrate (CAN)

Fertilizer grade ammonium nitrate (NH_4NO_3) contains 33-34% N. It has some disadvantages. It is quite hygroscopic and care must be taken to prevent caking and deterioration of its physical conditions during storage and handling. It is less effective for flooded soil than urea. It is more prone to leaching and denitrification than ammoniacal products. In India, use of NH_4NO_3 is prohibited because it is associated with some risk of fire or explosion.

In order to overcome the above problems associated with NH_4NO_3 fertilizer, it is usually mixed with limestone or lime ($CaCO_3$) and produced calcium ammonium nitrate (CAN) called ammonium nitrate-lime stone or calcium carbonitrate. It contains 25% N, 7-14% Ca and 4.5% Mg. Here, Ca comes from calcium carbonate (lime), and by virtue of Ca, CAN does not make the soil acidic. For further details, please see Section 2.11.

5.13.3. Single Superphosphate

This is the oldest fertilizer. It typically contains 16% P_2O_5, 9-12% S and 20% Ca. It's P and Ca comes from rock phosphate while S comes from H_2SO_4. It is the 2nd most important P-fertilizer after DAP and also the most important carrier of S in India. For further details, please see Section 3.14.

5.13.4. Potassium Sulphate (SOP)

Potassium sulphate contains 50% K_2O and 18% S, both in readily available form. The SOP has excellent physical properties and stores well even in damp condition (CRH = 96). It has a lower salt-index than muriate of potash (KCl) and thus does not add to the salinity hazard. For further details, please see Section 4.9.2.

5.13.5. Ammonium Phosphate Sulphate

It is a complex fertilizer containing ammonium sulphate and ammonium phosphate. It contains about 16 or 20% N, 20% P_2O_5 and 13-15% S. The S comes from $(NH_4)_2SO_4$. For further details, please see Section 3.18.

5.13.6. Gypsum ($CaSO_4.2H_2O$) and Phosphogypsum

Agricultural grade mineral gypsum contains 13-18% S and 16-19% Ca, while byproduct phosphogypsum contains 16% S and 21-29% Ca. Mineral gypsum comes mostly from mines in Rajasthan. Phosphogypsum is a byproduct of wet process of H_3PO_4 acid factories in which 4-5 tonnes of phosphogypsum is discharged for every tonne of P_2O_5. Gypsum is sparingly soluble in water (solubility <0.01 mol L^{-1}). It is not yet included in the FCO as fertilizer, but consider as amendment.

5.13.7. Magnesium Sulphate ($MgSO_4.7H_2O$)

This is the most commonly used fertilizer for correcting Mg-deficiency in soils. It is produced by reacting mineral magnesite with H_2SO_4.

$$\underset{\text{(Magnesite)}}{MgCO_3} + H_2SO_4 \rightarrow MgSO_4 \quad \text{...(5.3)}$$

It has 10% Mg and 12-13% S. It is a free-flowing crystalline salt. Being highly water soluble material, it can be used for fertigation and foliar application. Magnesium sulphate is available as: (i) Anhydrous magnesium sulphate ($MgSO_4$) containing 20% Mg and 26% S; (ii) Magnesium sulphate monohydrate ($MgSO_4.H_2O$), also known as Kiesserite, containing 16% Mg and 22% S; and (iii) Magnesium sulphate heptahydrate ($MgSO_4.7H_2O$), also known as Epsomite salt containing 9.6% Mg and 13% S.

5.13.8. Elemental Sulphur

Elemental sulphur (S) is primarily used as a raw material for the production of H_2SO_4 which in turn used for the manufacture of many fertilizers. The agricultural grade materials contain 80-100% S and directly used in alkaline-calcareous soils. Elemental S as such cannot be absorbed by plants and is converted to plant available SO_4^{2-} form in the soil by S-oxidizing bacteria such as *Thiobacillus*.

$$\text{Elemental S} \xrightarrow[\text{S-oxidizing bacteria}]{\text{Oxidation}} SO_4^{2-} \quad \ldots(5.4)$$

The availability of elemental S as fertilizer depends on the rate of oxidation which in turn depends on several factors such as (i) particle size of material, (ii) rate of application, (iii) methods of application, (iv) time of application, (v) sulphur oxidizing characteristics of soil, and (vi) environmental conditions.

The rate of oxidation increases with decrease in particle size. This is because grinding of material into fine powder form increases the surface area of the material, thereby, comes in contact with large volume of soil after application. When S is finely ground and mixed with soil possessing a high oxidizing capacity, it is usually just as effective as SO_4^{2-} sources. Time of application of elemental S should be at least 15 days before planting or sowing of crops. Elemental S requires some time for oxidation to convert SO_4^{2-} form and becomes available to plants. Thus, it cannot be used as top-dressing. It is placed near the root zone and mixed thoroughly with soil. Usually, high rates of elemental S may be required, because of slow conversion of this source to SO_4^{2-} form. The high rate of application is repeated only once in every 3 to 4 years.

5.13.9. Sulphur-Bentonite

Sulphur-bentonite is a product consisting of 90% elemental S and 10% bentonite, the latter being an expanding type of clay mineral which serves as a binder. It has been manufactured in USA and Canada since 1970. It is prepared by mixing bentonite with molten elemental S followed by cooling, solidifying, grinding and screening of material through 1-mm sieve.

Bentonite + molten S → Cooled → Solidified → Crushed → Screened → Product

Bentonite helps in absorbing more moisture and expands up to 20 times of its original size thus, helping in disintegration of elemental S to SO_4^{2-}-S. Because of the uncertainty of adequate availability of sulphur-bentonite during the initial growing season after application, it should be incorporated into soil at least 4-5 months before sowing/planting.

5.13.10. Potassium Schoenite

Potassium schoenite is manufactured by washing of the mineral langbeinite ($K_2SO_4.MgSO_4$) to remove the impurities. It is a mixture of sulphate of potash and sulphate of magnesium. It contains 23% K_2O, 10% MgO and 17% S. It is highly water soluble salt and can be used for fertigation and foliar spray.

5.13.11. Pyrites (Iron pyrites)

Pyrites or iron pyrites (FeS_2) are acidifying material and sedimentary geological origin. They are highly pyrophoric in nature and upon contact with air they are highly oxidizable. On addition to moist soil, pyrite gets oxidized by reacting with moisture and air to produce H_2SO_4 and ferrous sulphate as follows:

$$FeS_2 + H_2O + 3\frac{1}{2} O_2 \rightarrow H_2SO_4 + FeSO_4 \quad \text{... (5.5)}$$

Ferrous sulphate on hydrolysis forms H_2SO_4.

$$FeSO_4 + H_2O \rightarrow FeO + H_2SO_4 \quad \text{... (5.6)}$$

Thus, pyrites have marked acidulation effect leading to considerable solubilization. The first step in the oxidation process of pyrites is purely a chemical process. While, in second step, the reaction is believed to be facilitated by several bacteria belonging to genus *Thiobacillus (Thiobacillus oxidans)*. To get maximum benefit from the use of pyrites, optimal and favourable conditions for the auto-oxidation and hydrolysis of pyrites are a pre-requisite. Finer particle size of pyrites, aerobic and moist conditions and sodic environment favour continuous auto-oxidation. It contains about 22-24% S. It is advisable to apply pyrites in the field a few days before sowing of crops and mixed thoroughly with soil for better results.

5.13.12. Ammonium Thiosulphate

Ammonium thiosulphate [$(NH_4)_2S_2O_3$] is an inorganic ammonium salt composed of ammonium and thiosulphate ions in a 2:1 ratio. It is a white crystalline solid with ammonia odour, readily soluble in water, slightly soluble in acetone and insoluble in ethanol and diethyl ether. It contains 12% N and 26% S. It is the most popular S-containing product used in the liquid fertilizer industry. It is a non-corrosive fertilizer (liquid-fertilizer) and may be stored in mild steel or aluminum containers. This S-fertilizer is compatible with N-solution and NPK-liquid mixtures which are neutral to slightly acidic. Ammonium thiosulphate can be applied to the soil directly or in mixtures or it can be added to irrigation water, both sprinkler and open-ditch systems. When applied to soil, it decomposes to form colloidal elemental S and $(NH_4)_2SO_4$, the latter then converted to NH_4^+ and SO_4^{2-}. The NH_4^+ and SO_4^{2-} are available and taken up by plants. The colloidal S is converted to SO_4^{2-} by bacterial oxidation.

$$2\,NH_4SCN \xrightarrow{\text{decomposes}} (NH_4)_2SO_4 + \text{Colloidal S} \quad \ldots(5.7)$$
(Ammonium thiosulphate)

$$(NH_4)_2SO_4 \longrightarrow 2\,NH_4^+ + SO_4^{2-} \quad \ldots(5.8)$$

$$\text{Colloidal S} \xrightarrow[\text{Bacteria}]{\text{Oxidation}} SO_4^{2-} \quad \ldots(5.9)$$

5.13.13. Sulphates of Micronutrients

Sulphates salts of micronutrients, commonly used as micronutrients fertilizers in India are copper sulphate ($CuSO_4.5H_2O$), ferrous sulphate ($FeSO_4$), manganese sulphate ($MnSO_4$), zinc sulphate monohydrate ($ZnSO_4.H_2O$) and zinc sulphate heptahydrate ($ZnSO_4.7H_2O$). All these fertilizers are highly water soluble and dissociate into their respective cations when applied to soil along with SO_4^{2-} anion and taken by plants. Since sulphates of these micronutrients are water soluble, they behave similarly as in case of other water soluble S fertilizers.

5.13.14. Byproducts Containing Sulphur

Pressmud is a byproduct of sugar industry. It is also known as filter cake. During the sulphitation process of sugar, in order to remove the impurities like suspended particles, dispersed soil, bagasse, wax, fats, proteins, gums, pectins, tannins and inorganic salts such as Na and K present in the juice to get the sugar crystals, substantial amounts of lime (CaO) and SO_2 are used. Here, $CaSO_3$ is obtained along with other impurities in the form of sulphitation pressmud (SPM). It contains 1.0-1.5% N, 4-5% P_2O_5, 2-7% K_2O, 0.3-0.5% S and 3-5% lime.

5.14. Calcium Fertilizers

Calcium is not normally formulated as such into mixed fertilizers, but rather it is present as a component of the materials supplying other nutrients, particularly P-fertilizers. In case of SSP, it contains about 18-21% Ca; while, in TSP, it contains 12-14% Ca. Bone meal contains 23% Ca. The primary sources of Ca are the liming materials such as calcite, dolomite, *etc*. Gypsum is used as a fertilizer for long time for supplying Ca to groundnut. It is applied directly to soil at the early bloom stage of groundnut. Synthetic chelate such as Ca-EDTA contains 3-5% Ca and can be used as foliar spray.

Calcium nitrate [$Ca(NO_3)_2$] is a highly soluble fertilizer. The N is present in the form of NO_3^-. The fertilizer grade calcium nitrate contains 15.5% N and 19% Ca. Because of its high solubility, it is used in precision agriculture in poly-houses both for fertigation and foliar application. It is manufactured by treating calcite ($CaCO_3$) with nitric acid (HNO_3) as follows:

$$CaCO_3 + 2\ HNO_3 \rightarrow Ca(NO_3)_2 + H_2CO_3 \quad ...(5.10)$$

or, $$CaCO_3 + 2\ HNO_3 \rightarrow Ca(NO_3)_2 + H_2O + CO_2 \quad ...(5.11)$$

5.15. Magnesium Fertilizers

In general, use of Mg as fertilizers and soil amendments are low compared to Ca and S. The most important Mg containing fertilizer is potassium magnesium sulphate. It is a water soluble neutral salt containing 22% K_2O, 20-22% S, 7-11% Mg. It is now listed in the FCO. Magnesium sulphate as Epsomite salt ($MgSO_4.7H_2O$) is another source of both Mg and S. It is produced by reacting mineral magnesite with H_2SO_4. It contains about 10% Mg and 12% S. The other important sources are magnesia (MgO, 55% Mg), magnesium nitrate [$Mg(NO_3)_2$, 16% Mg), magnesium silicate (basic slag, 3-45% Mg), serpentine (26% Mg), dolomite ($CaCO_3$+$MgCO_3$, 12% Mg) and magnesium chloride [$MgCl_2.10H_2O$, 8-9% Mg].

Potassium magnesium sulphate, magnesium sulphate and magnesia are the most widely used materials in dry fertilizer formulations. While, magnesium sulphate ($MgSO_4$), magnesium chloride ($MgCl_2$) and synthetic chelated Mg are used in liquid fertilizer and foliar spray.

5.16. Fate of Secondary Fertilizers in Soils

Normally, secondary nutrients are not added to soil as fertilizer, but as amendments to reclaim sodic soils by using as gypsum ($CaSO_4.2H_2O$) or acid soils with liming materials like calcite ($CaCO_3$) or dolomite ($CaCO_3$ + $MgCO_3$). Secondary nutrients are also added to soils through other major fertilizers like ammonium sulphate, single superphosphate, potassium sulphate, *etc.*

Application of liming material neutralizes the soil acidity by increasing the availability of Ca^{2+} saturation on exchange sites as follows:

$$2\ H^+\text{-Exch} + CaCO_3 \longrightarrow Ca^{2+}\text{-Exch} + CO_2(\uparrow) + H_2O \quad ...(5.12)$$

$$2\ Al^{3+}\text{-Exch} + 3\ CaCO_3 + 6\ H_2O \longrightarrow 3Ca^{2+}\text{-Exch} + 2\ Al(OH)_3\,(\text{solid}) + 3CO_2(\uparrow) + 3H_2O \quad ...(5.13)$$

In case of alkali soils having pH of 8.5-10, application of gypsum helps to reduce the pH to 7.5-8.0 because of formation of calcite and releasing H^+ ions as follows:

$$Ca^{2+} + CO_2\,(g) + H_2O \rightleftharpoons CaCO_3\,(\text{calcite}) + 2\,H^+ \qquad ...(5.14)$$

Because of the above reaction, it keeps the soil colloids flocculated and predominantly saturated with Ca by displacing Na^+ ions which subsequently removed by leaching.

In case of sodic soils, gypsum and/or phosphogypsum, pyrites, and elemental S are usually used as amendments for reducing excess Na^+ ions on exchange complex by Ca^{2+} ions as follows:

$$CaSO_4 + 2\,Na^+\text{-Exch} \rightleftharpoons Ca^{2+}\text{-Exch} + 2\,Na_2SO_4 \qquad ...(5.15)$$

As a result of application of gypsum or liming in sodic/acid soils, the Ca, Mg and S content in soils improves by ameliorating the soil alkalinity and/or acidity. Deficiencies of Ca and Mg can be corrected by using dolomite ($CaCO_3$+$MgCO_3$).

Chapter 5

Micronutrients Fertilizers

Iron (Fe), zinc (Zn), copper (Cu), manganese (Mn), nickel (Ni), boron (B), molybdenum (Mo) and chloride (Cl) elements are known as micronutrients. The **micronutrients** are also called as **trace elements** or **minor elements** as they are required very small quantities by plants. The elements Fe, Zn, Cu, Mn and Ni are known as **micronutrient cations** and B, Mo and Cl are known as **micronutrient anions** and they behave accordingly in soils. Among these elements, Fe content in soil as well as plant is the highest and sometimes higher than phosphorus (P) and sulphur (S). In plant, the concentration of micronutrients often found up to 100 mg kg^{-1}, except that of Fe and Mn which may go up to 500 mg kg^{-1} (dry weight basis). Another element, cobalt (Co) is also regarded to be essential for the growth of certain higher plants and animals, and likely to be included in the list of micronutrients. The importance of micronutrients has been realized during the past four decades when micronutrient deficiencies were observed in most of the soils in our country, where intensive agriculture is practiced.

6.1. Soil Conditions Conducive to Micronutrient Deficiency

Improved instrumentation techniques and increased knowledge of mineral nutrition of plants help in the diagnosis of micronutrient deficiencies in soils and plants. The micronutrient deficiency in soils has been attributed to the following changes in Indian agriculture:

(i) Continuous removal of micronutrients from soil by high yielding crops like cereals, pulses and oilseeds.

(ii) Use of micronutrient-free high analysis fertilizers in modern agriculture. Intensive cultivated soils receiving large quantities of high analysis NPK fertilizers.

(iii) Light-textured soils with low pH, developed under very high rainfall (highly leached acidic sandy soils).

(iv) Soils with impeded drainage and high water table.

(v) Soils with low organic matter content.

(vi) Soils with very high pH (sodic soil) show Zn, Fe and Mn deficiency.

(vii) Organic soils (with very high content of OM such as peat and muck soils of Kerala).

(viii) Soils that have been over limed.

(ix) Calcareous and saline-alkali soils very high in pH (UP, Punjab, Bihar).

6.2. Status of Micronutrients in Indian Soils

With the exception of Zn, Cu and B, all other micronutrients are present in Indian soils in sufficient amounts to sustain agricultural productivity. The total content of Zn, Cu, B and Mo is comparatively low in light-textured soils of north India (Punjab, Haryana, Delhi, Uttar Pradesh and Uttarakhand and in sandy and alluvial soils of Western India (Gujarat). The total B content is low in eastern India, particularly in Bihar, Jharkhand, West Bengal, Assam and Odisha. Zinc-deficiency is found in soils having pH 8.0 or above and in coarse-textured soils. Zinc-deficiency is also noticed in almost all rice growing regions of the country. Boron-deficiency is found next to Zn in Indian soils - mostly in eastern states.

6.3. Sources of Micronutrients in Soils

The micronutrients are present in soils mainly as their oxides, sulphides and silicates form, and inherited from the soil-forming rocks and minerals (parent materials). Both deficiency and toxicity of micronutrients in soils are related to their concentration in the rocks and minerals from which the soils have developed. Besides inorganic minerals, organic matter also contributes substantially to the total micronutrients in surface soils. Amendments and fertilizer materials incorporated to soils also add substantial amount of micronutrient in soil.

Micronutrient cations exist mainly in the five fractions in soil. These are (i) water soluble, (ii) exchangeable, (iii) complexed, adsorbed and chelated, (iv) occluded and bound in carbonates, sesquioxides and secondary minerals, and (v) primary minerals.

Major natural sources of micronutrients present in soils are as follows:

- **Iron**: The sources of Fe are oxides, sulphides and silicates, Important Fe-containing minerals are Hematite (Fe_2O_3), Goethite (FeOOH), Magnetite (Fe_3O_4), Olivine [$(Mg,Fe)_2SiO_4$] and Pyrites (FeS_2).
- **Zinc:** It is present in the form of sulphide, oxides, carbonates and silicates. Important Zn-containing minerals are Sphalerite (ZnS), Smithsonite ($ZnCO_3$) and Hemimorphite [$Zn_4(OH)_2Si_2O_7.H_2O$].
- **Copper**: Copper is present in the form of sulphide and oxides and hydroxy carbonates. Major sources are Chalcocite (Cu_2S), Covellite (CuS), Chalcopyrite ($CuFeS_2$), Cuprite (Cu_2O), Tenorite (CuO), Azurite [$Cu_3(OH)_2(CO_3)_2$] and Malachite [$Cu_2(OH)_2CO_3$].
- **Manganese:** It is present in soils as oxides, carbonates and silicates. Important sources of Mn-minerals are Pyrolusite (MnO_2), Manganite (MnOOH), Rhodochrosite ($MnCO_3$), Hausmannite (Mn_3O_4) and Rhodonite ($MnSiO_3$).
- **Nickel:** It is present as oxides form. The important sources of Ni are Nickel ferrite ($NiFe_2O_4$), Nickel aluminate ($NiAl_2O_4$) and Nickel silicate (Ni_2SiO_4).
- **Boron:** Borates and borosilicates are the important forms of B present in soil. The major sources of B-mineral in soils are Borax ($Na_2B_4O_7.10H_2O$), Kernite ($Na_2B_4O_7.4H_2O$), Colemanite ($Ca_2B_6O_{11}.5H_2O$), Tourmaline [$(Na,Ca)(Mg,Fe)_3Al_6(BO_3)_3Si_6O_{18}(OH)_4$ and Kotoite [$Mg_3(BO_3)_2$].
- **Molybdenum:** Sulphides, oxides and molybdates are the major form of Mo in soils. Molybdenite (MoS_2), Ilsemannite ($Mo_3O_8.8H_2O$), Ferrimolybdite [$Fe_2(MoO_4)3.8H_2O$] and Powellite ($CaMoO_4$) are the major Mo-containing mineral in soil.
- **Chlorine:** Important sources of chlorides are Muriate of potash (KCl) and Sodium chloride (NaCl).

6.4. Physiological Functions of Micronutrients

(i) Zinc, Cu and Mn are required for carbohydrate and protein metabolism through several enzymatic activities.

(ii) Zinc is also required for synthesis of some growth promoting hormones.

(iii) Zinc-containing enzymes are carbonic anhydrase, alcoholic dehydrogenase, proteinases and peptidases.

(iv) Iron-containing enzymes are cytochrome oxidase, peroxidases and catalases.

(v) Copper-containing enzymes are cytochrome oxidase, ascorbic acid oxidase and polyphenol oxidase.

(vi) Manganese-containing enzymes are dicarboxylase and dehydrogenase. It also participates in photolysis of tricarboxylic acid (TCA) cycle or Krebs cycle. It is an essential element in photosystem II where it participates in photolysis.

(vii) Molybdenum is associated in nitrogenase and nitrate reductase and essential for N_2 fixation.

(viii) Cobalt is essential for symbiotic N_2-fixation and *Rhizobial* growth.

(ix) Boron is required to regulate the rate of water absorption and translocation of sugar in plants.

(x) Nickel is associated with several enzyme systems.

6.5. Micronutrient Deficiency Symptoms

(1) **Zinc deficiency:** Interveinal chlorosis, reduction of size of young leaves. *Khaira* disease of rice is reported in India. *Little leaf* in citrus, *white bud* of maize, *bronzing of leaves, rosetting of leaves, hens and chicks* disease of grapes.

(2) **Iron deficiency:** Interveinal chlorosis in young leaves with leaf margins and veins remaining green, *lime induced chlorosis*.

(3) **Copper deficiency:** Chlorosis of younger shoot tissues, white tips, necrosis, leaf distortion, *die-back symptoms*.

(4) **Manganese deficiency:** Interveinal chlorosis in dicotyledonous plants and greenish grey spots in monocotyledonous plants. The Mn-deficiency is commonly known as *grey speck* of oats, *marshy spot* of peas, *little leaf* in apple, *Pahala* blight of sugarcane.

(5) **Molybdenum deficiency:** Molybdenum deficiency resembles the N deficiency. It causes marginal scorching and rolling or cupping of leaves in cauliflower. *Yellow-spot* disease of citrus and *whip-tail* disease in cauliflower.

(6) **Boron deficiency:** Terminal buds and youngest leaves become discoloured. Internode become shorted and gives appearance of a bulky or a rosette appearance. Typical names given to B-deficiency in different crops are (i) *heart rot* of sugarbeet and mangold, (ii) *browning* or *hollow stem* of cauliflower, (iii) *top sickness* of tobacco, and (iv) *internal cork* of apple.

(7) **Chlorine deficiency:** Chlorosis in younger leaves and overall wilting due to possible effect on transpiration (coconut is sensitive to Cl-deficiency).

(8) **Nickel deficiency:** Characteristics deficiency symptoms of Ni have not yet been defined adequately.

6.6. Toxicity of Micronutrients

(i) In acid sulphate soils, micronutrient toxicity limits the crop growth under upland soil conditions. But under waterlogged situation in rice cultivation such problem disappears mainly due to rise in soil pH.

(ii) Iron-toxicity in rice is observed in soils rich in Fe, under highly reducing submerged soil condition due to presence of easily decomposable organic matter. Under such condition Fe-induced Zn-deficiency is also expected.

(iii) Soils surrounding mine effluents show Cu and Zn-toxicity.

(iv) Soil pH is very important factor which regulates the solubility and availability of micronutrients. A very low pH indicates that all micronutrient cations are in solution and may become toxic to common plants. In such cases one dominant micronutrient may hinder the absorption of other micronutrient cations. On the other hand, if pH is >7.0, micronutrients cations get precipitated on insoluble hydroxides. Under alkaline conditions, micronutrient cations are changed to their oxides or hydroxides of comparatively lower solubility.

Boron availability is affected by pH. At low pH, most of the B-compounds are soluble and thus B remains available to plants as H_3BO_3. In sandy soils having low pH, B is lost by leaching due to high rainfall. In fine-textured soils, B leaching is not a problem. At high pH, B availability decreases. Liming of soil reduces B availability due to lime-induced B-fixation by clay and other minerals as Na and Ca-borates are water soluble.

Molybdenum chemistry largely resembles that of P in soils. All soil factors which increase availability of P, enhances the availability of Mo. The Mo availability is very low in acid soils as in case of P. Molybdenum gets fixed in Fe and Al-compounds. The fixed Mo can be made available by application of P to soil which releases fixed Mo ions by anion exchange. Liming the acid soils increases Mo activity.

6.7. Antagonistic Effect

The following antagonistic effects have been well established on the uptake of micronutrients by crops:

(i) Excess of P adversely affects utilization of Zn, Cu and Fe.

(ii) Excess of Fe adversely affects utilization of Zn and Mn.

(iii) Excess of Zn, Mn and Cu induces Fe-deficiency in crops.

(iv) Excess of S and Cu induces Mo-deficiency.

(v) Excess of Fe causes lime induces deficiency of all micronutrients.

6.8. Synergistic/Beneficial Effects

(i) Beneficial effect of Mo and P in legumes.

(ii) Beneficial effect of Mn and Mo in field crops.

6.9. Micronutrient Management in Soil-Plant-Systems

Knowledge about factors affecting availability of micronutrients in soils is important for management of micronutrients.

(i) Due consideration is to be given to soil properties like texture, pH, soil moisture regimes, and $CaCO_3$ content.

(ii) Next consideration to be given is the intensity of cropping fertilizer consumption pattern. Higher cropping intensity and high dose of NP lead to multi-nutrient deficiencies.

(iii) The simplest management of micronutrient deficiency in soils is to add micronutrients fertilizers along with NPK in alternate years.

(iv) The micronutrient availability is regulated by soil moisture regime. The micronutrient availability is less in oxidized state than reduced state. Good drainage condition results in soil becoming aerobic and micronutrients are present in oxidized form. Thus, availability is less under such conditions.

6.10. Micronutrient Fertilizers

Both soluble and insoluble micronutrient carriers are utilized in routine agricultural practices to supply adequate level of micronutrients in soil. The common micronutrient fertilizers are given in Table 6.1.

6.10.1. Zinc Fertilizers

Among various inorganic sources, the most common Zn-containing fertilizer in India is zinc sulphate heptahydrate ($ZnSO_4.7H_2O$) which is found the most effective, frequently available, economically cheapest source of correcting Zn

Table 6.1. Common micronutrient fertilizers in India

Micronutrient	Nutrient materials	Formula	Nutrient content (%)
Zinc	Zinc sulphate heptahydrate	$ZnSO_4.7H_2O$	21% Zn, 15% S
	Zinc sulphate monohydrate	$ZnSO_4.H_2O$	33% Zn, 15% S
	Zinc oxysulphate	$ZnSO_4 + 4Zn(OH)_2$	55% Zn
	Zinc oxide	ZnO	55-70% Zn
	Zinc carbonate	$ZnCO_3$	52-56% Zn
	Zinc chloride	$ZnCl_2$	48-50% Zn
	Chelated Zn	Na_2Zn-EDTA, NaZn-NTA, Na_2Zn-HEDA	12% Zn
	Zincated urea	Zn-Urea	2% Zn, 43% N
	Zincated single superphosphate	Zn-SSP	0.5% Zn, 16% P_2O_5, 11.5% S
	Zincated phosphate suspension	Zn-phopshate	19.4% Zn, 12.9% P_2O_5
Copper	Copper sulphate pentahydrate	$CuSO_4.5H_2O$	24% Cu, 12% S
	Copper sulphate monohydrate	$CuSO_4.H_2O$	35% Cu, 12.8% S
	Copper oxysulphate	$CuSO_4 + 3Cu(OH)_2$	53-56% Cu
	Copper-EDTA	Cu-EDTA	9-13%
Iron	Ferrous sulphate heptahydrate	$FeSO_4.7H_2O$	19% Fe, 10.5% S
	Ferric sulphate	$Fe_2(SO_4)_3$	17% Fe
	Ferric chloride	$FeCl_3$	5-18% Fe
	Iron pyrites	FeS_2	22-24% S, 19% Fe
	Chelated Fe	Fe-EDTA	12% Fe
Manganese	Manganese sulphate trihydrate	$MnSO_4.3H_2O$	17% S, 26-28% Mn
	Manganese sulphate monohydrate	$MnSO_4.H_2O$	17% S, 30.5% Mn
	Manganese oxysulphate	$MnSO_4$.MnO	40-49% Mn, 6% S
	Manganese dioxide	MnO_2	55-65% Mn
	Manganese-EDTA	Mn-EDTA	5-12% Mn

Nickel	Nickel sulphate hexahydrate	$NiSO_4.6H_2O$	22% Ni, 12% S
Boron	Borax	$Na_2B_4O_7.10H_2O$	10.5% B
	Boric acid	H_3BO_3	17.5% B
	Disodium octa borate tetrahydrate	$Na_2B_8O_{13}.4H_2O$	20% B
	Disodium octa borate pentahydrate	$Na_2B_8O_{13}.5H_2O$	15% B
	Solubor	$Na_2B_4O_7.5H_2O$ + $Na_2B_{10}O_{16}.10H_2O$	19% B
Molybdenum	Ammonium molybdate	$(NH_4)_6Mo_7O_{24}.4H_2O$	52% Mo, 7% N
	Sodium molybdate	$Na_2MoO_4.2H_2O$	37-39% Mo
Chloride	Muriate of potash	KCl	48% Cl

deficiency in most of crops and diverse soils as compared to sparingly soluble Zn sources, chelates, and mixtures. Zinc sulphate monohydrate ($ZnSO_4.H_2O$) and zinc-oxysulphate are more popular in USA and other countries because these are more concentrated fertilizers than zinc sulphate heptahydrate, thereby reduce the storage, transport and application cost. Zinc sulphate heptahydrate ($ZnSO_4.7H_2O$) contains 21% Zn. While, zinc sulphate monohydrate ($ZnSO_4.H_2O$) contains 33% Zn. Zinc oxysulphate [$ZnSO_4$ and 4 $Zn(OH)_2$] contains 55% Zn and zinc oxide (ZnO) contains 55-70% Zn. The Zn-EDTA contains 12% Zn. The chelated form of Zn (Zn-EDTA) is free-flowing, crystalline material. It is evident that the use efficiency of EDTA-Zn is 4 to 5 fold higher than the $ZnSO_4$. However, chelated Zn has the disadvantage of high cost of 15 to 20 times higher than $ZnSO_4$. Other Zn-fertilizers are zincated urea (2% Zn), zincated phosphate (19.4% Zn).

Usually, Zn-fertilizer is applied as soil application @ 5-20 kg Zn ha^{-1} by broadcasting or band placement. Soil application of Zn is preferred method over less efficient foliar sprays. However, foliar spray with 0.5% $ZnSO_4$+0.25% lime suspension of 500 liter ha^{-1} is recommended for alleviating Zn deficiency, particularly for trees. For upland crops, normally lower rate is recommended for coarse textured sandy soils (5 kg Zn ha^{-1}) than heavy texture soils (10 kg Zn ha^{-1}). Application of 5-10 kg Zn ha^{-1} as $ZnSO_4.7H_2O$ by broadcasting is recommended just before the final ploughing followed by thorough mixing with soil helps in enhancing use efficiency of Zn by the crops besides crop yields. In case of chelated Zn such as disodium salt of Zn-EDTA (Na_2Zn-EDTA), NaZn-NTA and Na_2Zn-HEDTA), the rate of placement is 0.5-4 kg Zn ha^{-1}.

Synthetic Zn-EDTA chelates are found better than zinc sulphate in combating Zn deficiency in crops in non-calcareous loamy sandy soils but at par in calcareous and arid soils. However, their high cost than $ZnSO_4$ made chelated fertilizers most uneconomic and less effective for common use. Moreover, chelated Zn is easily translocated within the plants because unlike $ZnSO_4$ it is partly systemic. Zinc sulphate reduces P uptake and the P content of the shoots in plants, while Zn-EDTA increases it.

Micronutrient blended fertilizers sources such as zincated urea, zincated superphosphate and boronated superphosphate were found initially inferior to $ZnSO_4$ or Zn-chelates in highly Zn or B deficient soils as they mismatched with nutritional requirement of crops. But long-term effects of zincated urea, boronated superphosphate became as efficient as that of soluble sources in controlling hidden/emerging micronutrient deficiency in several crops.

Zinc oxide is effective for roots dipping and seed coating before transplanted crops than $ZnSO_4$. Zinc phosphate has been found efficient source for seed coating to control hidden hunger and found cheaper than ZnO.

Zinc sulphate heptahydrate, the most commonly used Zn-fertilizer when added to soil gets transformation and its Zn gets distributed into different fractions such as water soluble exchangeable and adsorbed organically-bound and acid soluble. The first three forms contribute to the available pool of Zn soils fertilized with high dose of P or limed reduce the availability of Zn.

6.10.2. Copper Fertilizers

The most important Cu-fertilizers are copper sulphate penta hydrate ($CuSO_4.5H_2O$) and copper sulphate monohydrate ($CuSO_4.H_2O$) containing 24% and 35% Cu, respectively. Other Cu-fertilizers are copper oxysulphate [$CuSO_4$ + $3Cu(OH)_2$] with 53-56% Cu and Cu–EDTA with 9-13% Cu.

6.10.3. Iron Fertilizers

The important Fe-containing fertilizers are (i) Ferrous sulphate ($FeSO_4.7H_2O$) with 19% Fe, (ii) Ferric sulphate ($Fe_2(SO_4)_3$] with 17% Fe, (iii) Ferric chloride ($FeCl_3$) with 5-18% Fe, and (iv) Fe-EDTA with 12% Fe. Iron fertilizers can be applied as soil application with higher dose of Fe (20-30 kg $FeSO_4$). However, it is less effective in alleviating Fe-deficiency. The higher use efficiency of Fe can be obtained by foliar application of Fe-salt with lesser dose (0.5-1.0% $FeSO_4.7H_2O$).

6.10.4. Manganese Fertilizers

The Mn-containing fertilizers are (i) Manganese sulphate trihyderate ($MnSO_4.3H_2O$) contains 26-28% Mn, (ii) Manganese sulphate monohydrate ($MnSO_4.H_2O$) contains 30.5% Mn, (iii) Manganese oxysulphate ($MnSO_4$ MnO) contains 40-49% Mn, (iv) Manganese dioxide (MnO_2) contains 55-65% Mn, and (v) Chelated-EDTA (Mn-EDTA) contains 5-12% Mn. As in case of Fe, foliar sprays of 0.5-1.0% of $MnSO_4$ solution are more effective in alleviating the Mn-deficiency in crops than soil application.

6.10.5. Boron Fertilizer

Major boron-containing fertilizers are: (i) Borax ($Na_2B_4O_7.10H_2O$) with 10.5% B, (ii) Boric acid (H_3BO_3) with 17.5% B, (iii) di-sodium octa borate tetrahydrate and di-sodium octa borate pentahydrate with 20 and 15% B, respectively, and (iv) Solubor [$Na_2B_4O_7.5H_2O + Na_2B_{10}O_{16}.10H_2O$] with 19% B. Usually, borax is applied as soil application, while boric acid (0.1-0.2% solution) is used as foliar spray and solubor is used as both soil application and foliar spray. Boron added to soil, after dissolution can get adsorbed by mineral soil constituents, or bound by organic matter, or remain as residual B apart from solution B.

6.10.6. Molybdenum Fertilizers

Sodium molybdate ($NaMoO_4.2H_2O$) with 37-39% Mo and ammonium molybdate [$(NH_4)_6Mo_7O_{24}.4H_2O$] with 52% Mo and 7% N are the two Mo-containing fertilizers used for alleviating Mo-deficiency in crops. When added to soils, ammonium molybdate undergoes transformation to form NH_4^+ and molybdate (MoO_4^{2-}). Similarly, when sodium molybdate is added, it form Na^+ and MoO_4^{2-}. The MoO_4^{2-} is taken by plants or it may form sparingly soluble complexes with Al^{3+} and Fe^{3+}. Soil application of 0.1-0.5 kg Mo ha^{-1} has been found to be satisfactory in correcting Mo-deficiency in crops. While, foliar spray of Mo (0.07-0.1% ammonium molybdate) solution is quite effective in alleviating Mo-deficiency. Seed treatment is also quite effective.

6.10.7. Chloride Fertilizer

Muriate of potassium (MOP) or potassium chloride (KCl) is the major Cl-containing fertilizer with 48% Cl. The other Cl-containing fertilizers used as fertilizers are NaCl and $CaCl_2$. Usually chloride is not applied as fertilizer supplier of Cl, except in coconut.

6.11. Form of Micronutrient Fertilizers

Micronutrient fertilizers can be grouped into four categories (i) inorganic salts, (ii) chelates, (iii) multinutrient formulations, and (iv) fortified fertilizers.

(i) **Inorganic salts** like $ZnSO_4$, $CuSO_4$, $FeSO_4$, $FeSO_4$, $MnSO_4$, *etc.* are readily water soluble and can be used both as soil application and foliar spray. These salts also contain sulphur along with the respective micronutrient. Fertilizers such as ammonium molybdate, borax, boric acid, *etc.* are also inorganic salts.

(ii) **Chelated form** is the next important form of micronutrient fertilizers. A chelate is a type of coordination compound in which a single metallic ion is attached by coordinated covalent bonds to a molecule or an ion called a ligand. The word "**Chelate**" is derived from the Greek word "*Chela*", meaning "crab's claw". It depicts how organic chelating agent, known as "ligand" holds the nutrient cation in its grip. Bonding in a chelate occurs because the ligand has at least two pairs of unshared electrons. Chelates are organo-metallic molecules of varying sizes and shapes in which the organic part binds the nutrient in a ring like structure. Common chelating agents used in chelated micronutrients are (i) EDTA (Ethylene diamine tetra acetic acid); (ii) HEDTA (Hydroxyethyle ethylene diamine triacetic acid); (iii) DTPA (Diethylene triamine penta acetic acid); (iv) EDDHA (Ethylene diamine dihydroxyphenyl acetic acid); and (v) NTA (Nitrilo-tri-acetic acid)

The chelated micronutrients when applied to soil become soluble but stable enough to remain as such for longer period till the limit is delivered to the plant root for uptake. It do not release its micronutrient part prematurely for re-fixation in soil. Stability of the metal-chelate bond affects the availability of the nutrients to plants. Plants can absorb these elements as cations and possibly also in chelated form from the soil solution or the leaf surface when applied through foliar spray. For efficient utilization of the chelate, it has to be broken at the root surface so that the cation enters the root and chelating agent stays in the soil (Mengel and Kirkby, 2001). The chelating agent thus works as a "courier service" from the point of chelation to crop uptake. Chelated micronutrients are more efficient but costly.

(iii) **Multinutrient formulations of micronutrients** are now-a-days gaining popularity because of multinutrient deficiencies in many parts of the country. At least five nutrients (N, P, K, S and Zn) are now have widespread practical importance and found deficient in Indian agriculture. Under such situations, multinutrient formulation performs better not only in terms of use efficiency but also their lesser cost involved due to storage, transportation and application.

(iv) **Fertilizers fortified** with micronutrients are a special class of fertilizers in which conventional fertilizers containing major nutrients are fortified with specific micronutrients. These are manufactured for three main purposes. These are (a) to increase value addition of conventional fertilizers such as urea, SSP, DAP, NPK Complexes, *etc.*; (b) to provide a much needed micronutrient along with major nutrients; and (c) to facilitate the uniform application of small amounts of micronutrient needed for crops without additional operation practices. For examples, urea fortified with Zn (Zincated urea containing 43% N and 2% Zn), zincated SSP (16% P_2O_5 and 0.4% Zn), boronated SSP (16% P_2O_5 and 0.18% B), NPK complex 10-26-26 fortified with 0.3% B or 0.5% Zn; NPK complex 12-32-16 fortified with 0.3% B or 0.5% Zn; DAP fortified with 0.3% B or 0.5% Zn.

Suspensions containing micronutrients like zincated phosphate (suspensions) has been included in the FCO. This product is used for treating seeds and other planting materials. It contains 12.9% P_2O_5 (total) of which 3.9% is citrate soluble and 19.4% Zn.

6.12. Methods of Application of Micronutrients

In general, for Mn, Cu and Fe, sulphate forms are generally used as fertilizer since they are water soluble form. Chelated-micronutrients are also water soluble and now becoming very popular for use in field and orchard crops. Micronutrients could be applied to the soils or directly to the crops in various ways. The common methods are:

(1) **Soil Application:** The required quantities are broadcasted or placed in furrows before sowing or planting. The material is mixed with some amount of dry soil or sand in order to increase the bulk for uniform application.

(2) **Foliar spray:** The green foliage of plants can absorb and utilize nutrients. Compare to soil application the doses of micronutrients for foliar sprays are low to avoid burning of leaves. Normally, 450-900 L ha^{-1} of micronutrient solution is applied. For crops in younger stages, require lesser amounts of solutions for spraying. While crops with more foliage or fruit trees like orange require more solutions for spraying.

(3) **Addition of micronutrients through mixed fertilizers**: The micronutrients essential for different regions of the country are added to the straight fertilizers or to fertilizer mixtures used in each locality. This enables a uniform spreading of the micronutrient without any additional cost of application. Phosphates mixed with B, Mo or Zn are commonly use in

western countries. In USA, granulated mixed fertilizers commonly contain micronutrients to enable the farmers to apply all the nutrients at a time are used. In India, now-a-days, customized fertilizers are being manufacturing by many fertilizer manufacturers which is specific to soil types, a specific crop based on its nutrient requirement, area/region of cultivation and season.

(4) **Seed soaking:** Small amounts of micronutrients are dissolved in water to soak the seeds for 12 hour before planting. In this method, it is very essential to use a relatively very low concentration of solution so that it do not adversely affect the seed germination.

(5) **Seed coating:** Micronutrients mixed with a small amount of soil are made into a paste which is then coated around the seeds and dried. These micronutrient coated seeds are then used for sowing.

Chapter 7

Slow-Release or Controlled-Release Fertilizers

Fertilizers have played a pivotal role in enhancing food grain production in India especially after the introduction of high yielding and fertilizer responsive crop varieties during the green revolution era. However, there are certain problems associated with fertilizers because of their relatively low nutrient use efficiency (NUE) or nutrient uptake by crops. The fertilizer use efficiency is 30-50% for N, 15-20% for P, 70-80% for K and 8-10% for S. In case of micronutrients, the use efficiency is only 1-2%. The low NUE is typically the result of higher release rate of fertilizers than the absorption rate by plants, and/or transformation of fertilizers/ nutrients to forms that are not bioavailable to crops.

Application of N and P fertilizers at levels exceeding plant requirements due to low acquisition efficiency leads to significant environmental consequences in many parts of the world due to losses, such as nitrate (NO_3^-) leaching, NH_3 volatilization, and nitrous oxide (N_2O) emission. Excessive use of fertilizer N affects the groundwater. Similarly, excessive use of N and P fertilizers leads to surface runoff which in turn causes eutrophication in aquatic ecosystems. These negative environmental consequences associated with fertilizer inputs further emphasize the need of technological approaches to improve nutrient management in modern agriculture. As such, there is great need in developing innovative fertilizers to minimize the losses of nutrients and increase their efficiencies.

In addition, current agricultural activities contribute up to 20-30% to the annual atmospheric emissions of greenhouse gases (GHGs), such as methane (CH_4) and carbon dioxide (CO_2). Even higher contribution is being noted for N_2O (about 60%), which is a potent GHG and catalyst for stratospheric ozone depletion, with more than 300 times the global warming potential than CO_2. Its emission is closely related to mineral fertilizer input. High fertilization rates lead to N and P losses with negative impacts on atmospheric GHGs concentrations and water quality.

There is an urgent need to improve NUE in agricultural systems, and to manage biogeochemical cycles in a sustainable way. This includes the development and application of modern technology as alternatives to conventional fertilization. There is a need to reduce the use of fertilizers/nutrients in agricultural systems by increasing NUE. The recent development in smart fertilizer delivery systems including nanofertilizers or the new generation fertilizers is a good alternative to commercially available fertilizers to improve the NUE and enhance resource utilization. Smart fertilizers may be a solution to enhance food production and environmental quality.

7.1. Mechanisms of Losses of N from Soils

Ammoniacal-N (NH_4^+-N) applied through commercial fertilizers like ammonium sulphate or as intermediate product of urea or produced by mineralization of organic matter (ammonification process) undergo conversion to nitrate-N (NO_3^--N) by the process known as **nitrification**. The term nitrification refers to the conversion of NH_4^+ to NO_3^-. This process is brought about by the nitrifying bacteria, which are a specialized group of soil microorganisms gaining their energy by oxidizing NH_4^+, while using CO_2 as their source of carbon to synthesize organic compounds. Nitrification is a two-step process, wherein in the first step *Nitrosomonas* converts NH_4^+ to NO_2^- and in the second step occurs through *Nitrobacter* species, which convert NO_2^- to NO_3^- as follows:

$$NH_4^+ \xrightarrow[\text{Nitrosomonas}]{+O_2} NO_2^- \xrightarrow[\text{Nitrobactor}]{+O_2} NO_3^- \quad \ldots (7.1)$$

The NO_3^- form of N is completely mobile and moves largely with the soil water either horizontally or vertically. Under conditions of excessive rainfall or irrigation, it is leached out from the roof zone or lost as run-off. As a result, considerable losses of N occur by leaching and subsequently by denitrification from soils. Denitrification refers to the process in which NO_3^- and NO_2^- are converted to gaseous compounds (NO, N_2O and N_2) by soil microbes under conditions of low oxygen supply. When soils become waterlogged, O_2 is excluded and anaerobic decomposition takes place. Some anaerobic organisms have the ability to obtain their nitrates and release elemental N_2 and nitrous oxide. The sequences usually involve during denitrification steps are shown as:

$$2\,HNO_3 \xrightarrow[-2H_2O]{+4H} 2\,HNO_2 \xrightarrow[-H_2O]{+2H} 2\,NO \xrightarrow[-2H_2O]{+2H} N_2O \xrightarrow[-H_2O]{+2H} N_2 (\uparrow) \quad \ldots (7.2)$$

Thus, a considerable amount of applied N is lost by leaching and denitrification. Using ^{15}N tracer technique, it has been found that the gaseous N losses from wetland rice fields *via* nitrification-denitrification range from 42-51% of applied N. However, the losses under upland conditions are quiet low and range from 10-20% of the applied fertilizer N.

Another mechanism of losses of applied N is by ammonia (NH_3) volatilization to the atmosphere. Ammonia volatilization occurs when NH_4^+ in the soil is converted to NH_3 at pH above 7.0, which is then lost as a gas. Significant losses of NH_3 occur from surface-applied N-fertilizers and organic manures containing NH_4^+-N or urea. The amount of total N loss from fertilizers containing urea due to NH_3 volatilization can vary considerably from 0 to 50% of the applied N. Typical losses from urea broadcast, without rain/irrigation for at least a week after application, may range from 10 to 20% of the applied N. Ammonia volatilization loss is maximum with urea and anhydrous ammonia, intermediate with urea ammonium nitrate (UAN) solution, and the least with ammonium nitrate (NH_4NO_3), ammonium chloride (NH_4Cl) and ammonium sulphate [$(NH_4)_2SO_4$].

As a consequence, crop recoveries of N seldom exceed 50-60% of that of the added fertilizers. These are of great importance now-a-days because of environmental concerns over the excessive movement of NO_3^- into surface and groundwater, as well as the effects of gaseous N losses on the upper atmosphere. The NO_3^- lost by leaching or transported in surface runoff results in reduced soil fertility, causes eutrophication of surface waters and increases concentrations in drinking-water. High NO_3^- levels (>10 ppm) in drinking water are associated with *Methemoglobinemia* or the blue baby syndrome in infants.

There has been ongoing search for alternative N-fertilizers with greater efficiency. One means of increasing the efficiency of applied fertilizers is to split application, particularly in light textured soils and for long duration crops. The alternative means of increasing the efficiency is by developing some new formulations or novel or ideal source of N-fertilizers.

An ideal source of N would be the one that would release N at a rate approximately equal to that of the plant needs *i.e.* uptake by plants. Thus, whatever the amount of N removed from the soil by plant uptake may be replenished by releasing that amount only from the fertilizer source. The N-fertilizers with slow availability have been developed, so that they supply N continuously over an extended period, thereby increasing the efficiency of fertilizer. These fertilizers can be known as slow-release fertilizers (SRFs) or controlled release fertilizers (CRFs).

7.2. Slow-Release Fertilizer or Controlled-Release Fertilizer

Slow-release fertilizer (SRF) or controlled-release fertilizer (CRF) may be defined as a fertilizer containing a plant nutrient in a form which either delays its availability for plant uptake and use after application, or extends its availability to the plant significantly longer than a reference rapidly available nutrient fertilizer such as ammonium nitrate (NH_4NO_3) or urea (NH_2CONH_2), monoammonium phosphate ($NH_4H_2PO_4$) or diammonium phosphate [$(NH_4)_2HPO_4$] or potassium chloride (KCl). The SRFs or CRFs are granulated fertilizer that releases nutrients gradually into the soil and provide a steady supply of plant nutrients over an extended period of time. They contain plant nutrient(s) in a form that delays their initial availability. The slowness of the release is determined by the low solubility of the chemical compounds in the soil moisture.

7.2.1. Difference between SRFs and CRFs

Both the terms of SRFs and CRFs are used to define the fertilizer materials that reduce the solubility so as to supply the nutrients to crops for a longer period. There is no official differentiation between SRFs and CRFs.

However, in strict sense, the term CRF is applied to those fertilizers in which the factors controlling the rate, pattern and duration of release are ***well known and controllable*** during their preparation. The CRFs are typically coated or encapsulated with inorganic or organic materials that control the rate, pattern, and duration of plant nutrient release. Most importantly, the release rate of a CRF is designed in synchronization to meet the changing crop nutrient requirements. The CRFs provide the plant with available nutrients for a longer time compared to normal fertilizers, such as urea. The products, such as coated or encapsulated products are commonly referred to as CRFs.

The SRFs are those fertilizers which involve a slower release rate of nutrients than conventional water-soluble fertilizers and CRFs, but the rate, pattern, and duration of release are ***not controllable*** because they depend on microbial decomposition/degradation whose effectiveness is dependent on temperature and moisture conditions. The microbially decomposed N products, such as urea-formaldehydes, are commonly referred to as SRFs.

The bioavailability of SRFs/CRFs is significantly longer than normal fertilizers such as ammonium nitrate and urea. The delay of initial availability or extended time of continued availability may occur by a variety of mechanisms. These include controlled water solubility of the material by semi-permeable coatings, occlusion, chemical forms or by slow hydrolysis of water-soluble low molecular weight compounds.

7.2.2. Functions of Ideal SRFs/CRFs

The SRFs/CRFs are developed mainly due to increase the NUE by crops. A SRF/CRF ideally allow the growing plant to improve nutrient uptake or nutrient recovery by:

(i) Releasing its nutrient slowly that synchronizes the plant demand throughout the growth of crops;

(ii) Reducing nutrient losses via leaching and runoff;

(iii) Reduction of chemical and biological immobilization reactions in soils which cause plant-unavailable forms; and

(iv) Reduction of rapid nitrification and N losses through NH_3 volatilization and denitrification.

(v) The SRFs and/or CRFs hold the nutrient within the fertilizer granule until plants actually require it. Thus, these can reduce nutrients losses and enhance NUE, decrease the labour costs and protect the environment.

Other advantages of SRFs are reduction of seed germination and seedling damage due to high concentration of salts, reduction of leaf burn from heavy rates of surface-applied fertilizers, improved crop quality, improved storage and handling properties.

Since conventional fertilizers are soluble in water, the nutrients can disperse quickly as the fertilizer dissolves. Because SRFs/CRFs are not water soluble, their nutrients disperse into the soil more slowly. The fertilizer granules may have an insoluble substrate or a semi-permeable membrane or coating or jacket that prevents dissolution while allowing nutrients to flow outward.

In order to make fertilizers more efficient, several options are available such as slowing their release, inhibiting conversion to forms that are less stable in the soil, or enhancing availability to plants. These include chemical additives, biological inhibitors, and coatings that physically constrain nutrient availability in the soil. Use of these additives depends upon potential benefits in relation to the cost, considering appropriate agronomic, economic, and environmental factors. Application of SRFs and CRFs can potentially decrease fertilizer use by 20-30% of the recommended rate of a conventional fertilizer while obtaining the same yield. Compared to the large amount of fertilizers used throughout the world, the total use of SRFs and CRFs is still small and comprises only about 0.15% of the total use of nutrients. These are being expensive than conventional fertilizers due to the additional labour involved and equipment required to produce coated fertilizers and the cost of the coating materials.

7.2.3. Classifications of SRFs/CRFs

According to the mode of release, the SRFs or CRFs can be grouped into following four types:

(1) Materials of low water solubility that must undergo chemical and/or microbial decomposition (*e.g.* Urea-formaldehyde, Crotonylidene Diurea, Isobutylidene diurea, Urea-Z, Oxamide, *etc.*);

(2) Conventional water soluble products treated to impede dissolution by physical barrier *i.e.* by a coating (*e.g.* Coated fertilizers like S-coated, Neem cake coated, Lac coated, Polymer coated, *etc.*);

(3) Sparingly soluble compounds or materials of limited water solubility and plant available forms (*e.g.* metal ammonium phosphate); and

(4) Soluble or relatively water-soluble substances that gradually decompose (*e.g.* Guanylurea sulphate, and guanylurea phosphate)

7.2.3.1. Materials of low water solubility that must undergo chemical and/or microbial decomposition

This group of SRF is comprised of chemical compounds that are only slightly soluble in water or in the soil solution. The rate of liberation of N from them is related to their water solubility and to the rate of microbial action and chemical hydrolysis, which in turn related to particle size and other factors. The important SRFs under this category are discussed below:

(a) Urea Formaldehyde

The condensation of urea with aldehyde is one of the most common methods for preparing SRFs. Urea reacts with a number of aldehydes to form sparingly soluble compounds containing over 30% N which decompose slowly in soil by chemical and/or biological action. One of the best known products in this category is **"urea-formaldehyde"**, often referred to as "**urea-form**". Urea aldehyde compounds are the major representatives of this group that are commercially available.

They are white, odorless solids having 38% N, which are made by condensing urea with formaldehyde. It is the first major synthetic source of N developed for controlled availability. Urea-form instead of definite compound consists of methylene urea polymers varying in chain length and molecular weight. The availability of N of the product depends on the mole ratio of urea to formaldehyde. That means the availability will decrease with increase in mole ratio of urea to formaldehyde, resulting a product which is completely insoluble.

The principal reactions are given below:

$$\underset{\text{(Urea)}}{NH_2CONH_2} + \underset{\text{(Formaldehyde)}}{CH_2O} \xrightarrow{\text{Acid catalyst}} \underset{\text{(Monomethylol urea)}}{NH_2CONH\text{-}CH_2OH} \quad ...(7.3)$$

$$\underset{\text{(Urea)}}{NH_2CONH_2} + \underset{\text{(Monomethylol urea)}}{NH_2CO\text{-}NH\text{-}CH_2OH} \xrightarrow[\text{-}H_2O]{\text{Acid catalyst}} \underset{\text{(Methylene diurea)}}{NH_2\text{-}CO\text{-}NH\text{-}CH_2\text{-}NH\text{-}CO\text{-}NH_2} \quad ...(7.4)$$

The above reaction can also be represented as:

$$\text{or, } U + U\text{-}CH_2\text{-}OH \xrightarrow[\text{-}H_2O]{\text{Acid catalyst}} U\text{-}C\text{-}U \quad ...(7.5)$$

$$\underset{\text{(Methylene diurea)}}{NH_2\text{-}CO\text{-}NH\text{-}CH_2\text{-}NH\text{-}CO\text{-}NH_2} + \underset{\text{(Monomethylol urea)}}{NH_2CONH\text{-}CH_2OH} \xrightarrow[\text{-}H_2O]{\text{Acid catalyst}}$$

$$\underset{\text{(Dimethylene triurea)}}{NH_2CONH\text{-}CH_2\text{-}NHCONH\text{-}CH_2\text{-}NHCONH_2} \quad ...(7.6)$$

$$\text{or, } \underset{\text{(Methylene diurea)}}{U\text{-}C\text{-}U} + \underset{\text{(Monomethylol urea)}}{U\text{-}CH_2\text{-}OH} \xrightarrow[\text{- }H_2O]{\text{Acid catalyst}} \underset{\text{(Dimethylene triurea)}}{U\text{-}C\text{-}U\text{-}C\text{-}U} \quad ...(7.7)$$

$$\underset{\text{(Dimethylene triurea)}}{U\text{-}C\text{-}U\text{-}C\text{-}U} + \underset{\text{(Monomethylol) urea}}{U\text{-}CH_2OH} \xrightarrow[\text{-}H_2O]{\text{Acid catalyst}} \underset{\text{(Trimethylene tetra urea)}}{U\text{-}C\text{-}U\text{-}C\text{-}U\text{-}C\text{-}U} \quad ...(7.8)$$

In reaction 6.3, one mole of urea reacts with one mole of formaldehyde to form monomethylol urea. Here, one methylene radical ($-CH_2-$) attaches 2 urea radical (NH_2CONH-). The compound monomethylol urea is most soluble, but it is less soluble than urea. Solubility goes on decreasing as reaction proceeds. As the length of this reaction increases the solubility of the resultant product decreases in the order:

$$\xrightarrow[\text{Solubility decreases}]{\text{Diurea > Triurea > Tetraurea}}$$

But during reaction, it is not possible to get a pure compound rather a mixture of so many compounds may be obtained. Thus, solubility is average of all compounds.

A typical urea-form may contain 30% of its N in forms that are soluble in cold water (25 °C). The cold water N fraction nitrifies almost as quickly as urea. Solubility in hot boiling water is a measure of the quality of the remaining 70% of its N. At least 40% of the N insoluble in cold water should be soluble in hot water for acceptable agronomic response.

The SRFs are characterized by their **activity index (AI)** which is used by the **Association of Official Agricultural Chemists (AOAC)** to evaluate the suitability of urea-formaldehyde compounds or urea-form or SRFs.

The activity index is defined as:

$$\text{Activity index (AI)} = \frac{(\%\ \text{CWIN} - \%\ \text{HWIN})}{\%\ \text{CWIN}} \times 100 \quad \ldots(7.9)$$

where, % CWIN = % cold water (25 °C) insoluble N; % HWIN = % hot water (98-100 °C) insoluble N.

Here,

(i) % CWIN > % HWIN

(ii) (% CWIN - % HWIN)/ % CWIN = Fraction of cold water insoluble N soluble in hot water.

(iii) (% CWIN - % HWIN) = Amount of N that is soluble in hot water.

(iv) The AI provides an estimate of the fraction of relatively long-lasting release of N from SRFs by about six months.

(v) The AI value of CRFs should be ≥ 40. As a result of technological developments, a shift has occurred over the past decade in AI values from 40% to 55-60%.

The suitability of these compounds as fertilizers depends on:

(a) The quantity of CWIN which is the source of the slowly available N.

(b) The quality of CWIN as determined by its AI, which reflects the rate at which the CWIN will become available.

The application of urea-form as a source of N has been reported to give satisfactory performance when applied to plants having steady and prolonged period of N demand. Since their mineralization rate is slow, they cannot be used for short duration crops, because of their immediate needs of N for quick growth can not be met. For long duration crops such as sugarcane, forest plantation, grasses, these fertilizers are best suitable. These materials can be best utilized by preparing a mixture of urea: urea-form at 50:50 ratio, and applying to soil. Here, the soluble fraction *i.e.* urea will meet the initial requirement of N and urea-form will meet the N requirement at the latter stage of crop growth.

(b) Crotonylidene Diurea (CDU)

This slow-acting N compound is formed by the reaction of urea with crotonaldehyde or acetaldehyde by the condensation of 2 moles of acetaldehyde and 2 moles of urea. Powdered CDU containing 30% N has been used directly as a fertilizer. This fertilizer was first developed in Germany in 1957 and marketed under the name of **Floramid** which is a granular product with N content of 28% and composed of 90% CDU and 10% NO_3^--N. The product decomposes slowly in the soil and the N requirement extends over a longer period or growing season. It is mainly used on professional turf, in nurseries, greenhouses, on lawns and for gardens and landscaping, and in high-value crops. However it has yet been proved to be sufficiently effective to the field crops probably owing to the fact that they are decomposed by soil bacteria.

(c) Isobutylidene Diurea (IBDU)

Isobutylidene diurea (IBDU) is an organic compound with formula $(CH_3)_2$-CH-CH$(NHCONH_2)_2$. It is a derivative of urea molecule (NH_2CONH_2). It functions as a SRF owing to its low solubility, which limits the rate of its hydrolysis to urea. It is produced commercially in Japan. It is produced by the condensation reaction of urea and isobutyraldehyde in a 2:1 mole ratio.

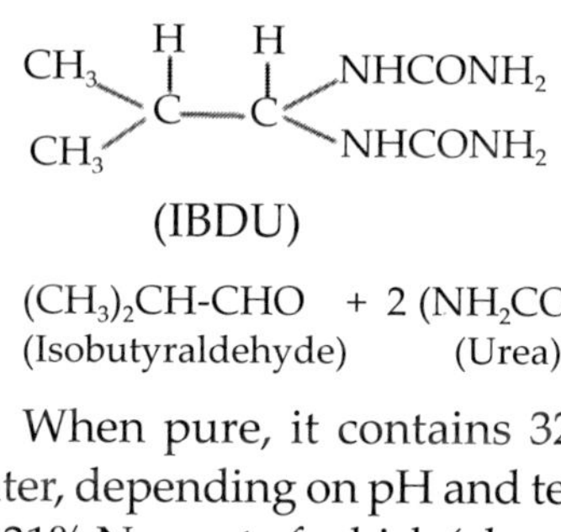

(IBDU)

$$(CH_3)_2CH\text{-}CHO + 2\,(NH_2CONH_2) \rightarrow (CH_3)_2\text{-}CH\text{-}CH(NHCONH_2)_2 + H_2O \quad ...(7.10)$$

(Isobutyraldehyde) (Urea) (Isobutylidene Diurea, IBDU)

When pure, it contains 32% N and has a solubility of only 0.1-0.01% in water, depending on pH and temperature. Fertilizer-grade IBDU contains about 30-31% N, most of which (about 90%) is water insoluble. It can also be granulated with soluble NPK sources to produce a range of compound fertilizers. Its mineralization in soil is accelerated when pH, temperature and moisture conditions are favourable for plant growth.

(d) Urea-Z

Urea-Z is formed by the reaction of urea and acetaldehyde, yielding a mixture of ethylene diurea and diethylene, together with some unreacted urea. The study of this substance started in Germany in 1956. It contains about 33-38% N, of which about 15% is initially water soluble and the remainder is slowly soluble. Its decomposition and release of N in soil is more rapid than most of the common slow-release N materials.

(e) Oxamide (NH_2-CO-CO-NH_2)

Oxamide (NH_2-CO-CO-NH_2) is another slow release N-fertilizer containing 31-8% N. The availability of N decreases as the particle size increases. Oxamide is a derivative of di-amide of oxalic acid, first reported in the late 1950 by Japanese workers. It is a white non-hygroscopic compound. Oxamide is practically insoluble in cold water and at 25 °C its solubility in only 0.4% by wt. Following dissolution, hydrolysis occurs primarily by means of microbial cleavage of the C-C bond, resulting in the formation of ammonium carbonate [$(NH_4)_2CO_3$]. Powdered and fine particles of oxamide nitrify faster than ammonium sulphate and at a rate similar to urea in acid soils. Presently it is not produced because of high cost of production. It is used for turf grasses and horticultural crops.

7.2.3.2. Conventional water soluble products treated to impede dissolution by physical barrier *i.e.* by a coating (Coated fertilizers)

Soluble N-fertilizers may be protected against losses due to leaching and denitrification by a coating that acts as a physical barrier. These coatings or membranes are rather inert and water resistant. A wide range of natural or synthetic polymers (*e.g.* polyurethane, polyethylene, *etc.*), waxes, paraffins, asphaltic compounds and elemental sulphur can successfully control the rate of release.

In general, the coatings can be of 3 types:

(i) **Semipermeable membrane** where water diffuses through the membrane (Fig. 7.1) and dissolves the core of water soluble N, creating an increase in osmotic pressure which forces the dissolved N out through the coating into the soil N. Example of this type of coating material is synthetic polymers.

(ii) **Perforated impermeable** membrane, where pinholes through the membrane coatings (Fig. 7.2) permit outward passage of water soluble N.

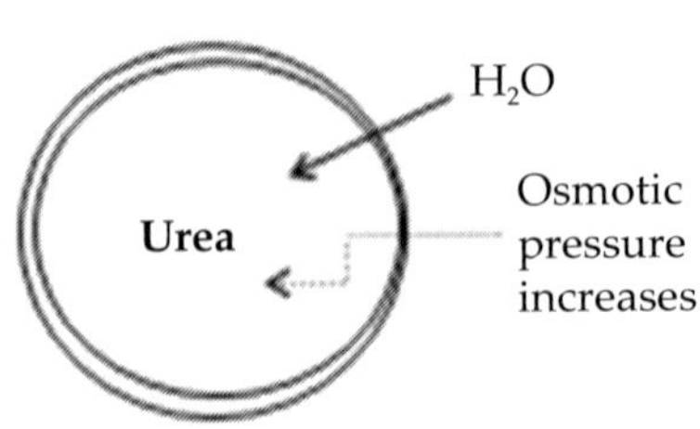

Fig. 7.1. Semipermeable membrane

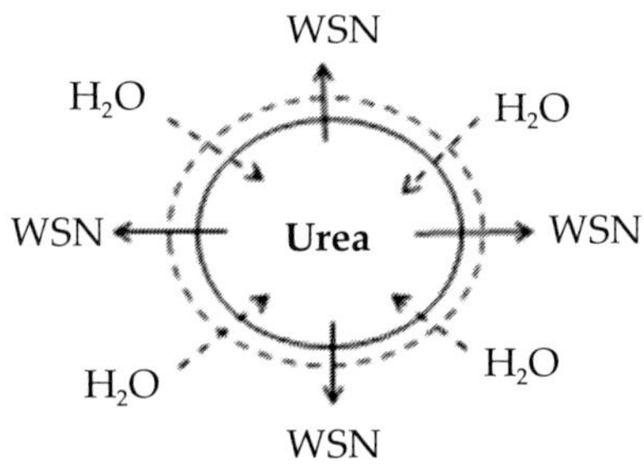

Fig. 7.2. Perforated impermeable membrane

(iii)Solid impermeable membrane comprised of materials that are degraded by soil microorganisms, thus creating openings through which dissolved N can be liberated (Fig. 7.3). Examples are elemental S, gypsum, *etc.*

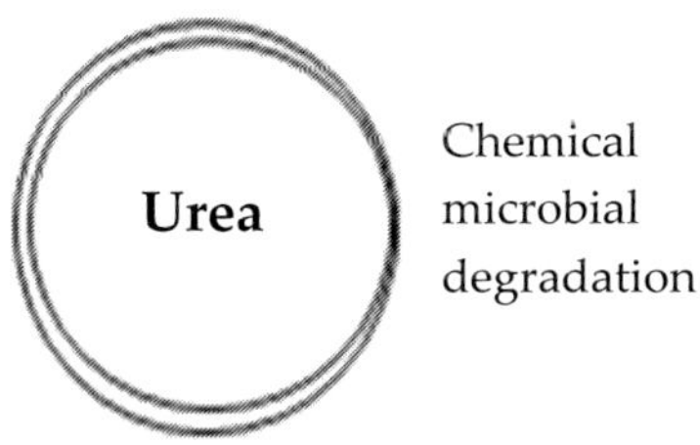

Fig. 7.3. Solid impermeable membrane

Coatings may act only as a barrier or they may also be a source of plant nutrients. The selected coatings should be thin otherwise it will increase the particle size of fertilizer and lower the nutrient content. At the same time coating should not be very thin, then the rate of release of N will be fast and slow release property will be defeated.

(a) Sulphur-Coated Urea (SCU)

Sulphur coated urea (SCU) is a controlled release N-fertilizer prepared by coating urea granule with sulphur which gives an impermeable layer that slowly degrades through microbial, chemical and physical processes. Sulphur is the most common substance used as coating material to protect urea granules from fast dissolution. It was first prepared in 1961 by the **Tennessee Valley Authority (TVA)**.

The basic process involves formation of a thin S-shell around each urea particle. This is done by spraying molten S on the falling urea particles in a rotating drum. In a second drum, molten sealant (a mixture of polyethylene and bright stock oil) is applied to seal the microscopic pores and cracks with the S-coatings. The SCU contains 36-38% N and 16% S. Here, N comes from urea and S from elemental S. The amount of the coating controls the time and rates at which urea leaves the granule and subsequently undergoes transformation such as hydrolysis and nitrification.

The SCU has the greatest potential for use in situations where multiple application of soluble N sources are needed during the growing seasons, particularly on sandy soils under high rainfall or irrigation. It is advantageous for use on sugarcane, pineapple, grass forages, turf, ornamentals, fruits and rice with intermittent flooding. Another advantage of SCU is its S content. Although elemental S in coating may not be available in the first year after application to correct deficiencies, but it can be an important source of plant available S in succeeding years. More is the coating size, less is the release rate. Here, S acts as a physical barrier. The release of N increases with increase in temperature. The rate of release also depends on the activity of microorganisms and their population in soil. Increasing moisture level, increases the release rate. Due to the simplicity and relatively low cost of S as a coating material, SCU has become the most commonly used coated-urea product in many parts of the world.

Although SCU has many advantages, there are several disadvantages:

(i) The SCU has proved highly efficient source of N, particularly in rice, but it is not popular in India due to high cost of S. Moreover, there is no good source of S in India and most of it is imported from other countries, thus increases cost of production.

(ii) The SCU is vulnerable to mechanical impact or abrasion. Excessive impact will break down the coatings and alter or destroy the slow-release characteristics of the product.

(iii) It is more slow release N-fertilizer. Thus, the availability of N is very less. For dissolution of S-coating, bacterial activities are necessary.

(iv) The SCU needs proper manufacturing attention for proper thickness which increases the cost of production.

Besides sulphur coated urea, there are other materials used for coating urea granules. These are plastic coated, wax coated, coaltar coated, lac coated, gypsum coated, and rock phosphate coated urea.

In case of slow-release or coated urea, the entire doses of N should be applied at basal application.

(b) Coated Calcium Carbide (CCC)

Encapsulated calcium carbide (CaC_2) pellets are mixed with fertilizer to produce a controlled release fertilizer known as coated calcium carbide (CCC). Due to its rapid reaction with water, it is mostly applied to soil in some encapsulated form so that a sustained supply of acetylene (C_2H_2) gas may be produced to inhibit nitrification or the activity of ammonium oxidizing enzyme for a longer period. For this purpose, waxes are used for the encapsulation of calcium carbide (Banerjee and Mosier, 1989). The calcium carbide particles are successfully coated with waxes (1:1). The other materials used for this purpose are enamel black paint, gelatine medical capsules, or polyethylene as hydrophobic material in calcium carbide formulations. To coat calcium carbide with these coating agents, calcium carbide particles (2-4 mm diameter) are mixed thoroughly with molten waxes or paint mechanically in a drum shape container. Wax and paint coated calcium carbide is then mixed with plaster of paris to keep particles separated from each other.

On contact with water these pellets slowly disintegrate and C_2H_2 is released, which has proved effective in inhibiting the nitrification of NH_4^+-based fertilizers as follows:

$$CaC_2 + 2\ H_2O = C_2H_2 + Ca(OH)_2 \quad \text{...(7.11)}$$

It is reported that C_2H_2 inhibits the enzyme, ammonia mono-oxygenase, which is responsible for the oxidation of NH_4^+ to NO_3^-. Acetylene in the soil not only inhibits nitrification, but its reduction to ethylene (C_2H_4) by soil microbes having nitrogenase activity is also beneficial because ethylene is a plant growth hormone and is involved in a number of physiological processes during plant growth and development. In case of cereals, ethylene has been reported to reduce lodging and enhance grain yield. Although they served the purpose well but under high temperature conditions and large scale soil application made this type of formulation ineffective or impractical.

(c) Polymer Coated Urea (PCU)

Polymer coated urea (PCU) can be prepared by coating urea with polymers such as resins or mineral-based coating agents. These polymers act as semi-permeable membranes or impermeable membranes with tiny pores having hydrophobic or water insoluble property. Upon application of urea to soil, water enters into the urea granule through coating to dissolve urea and N diffuses out through porous polymer membrane. These coatings can protect the fertilizer particles from a few weeks to several months depending on the coating and the conditions in the soil. These polymer coated products release nutrients in synchronization with crop demand which is taken by plants thereby increase nutrient use efficiency.

(d) Urea Super Granule (USG)

Urea super granule (USG) is a special type of urea having spherical granule size with ~1 cm diameter granules or pellets and 1 g weight produced primarily to facilitate point placement of N in flooded rice soils. The materials are presumably produced by briquetting or compacting urea prills. Usually, they are placed in deep water rice cultivation. Use of USG by rice farmers is becoming popular in Bangladesh. The USG outperforms commercial urea prills in rice on most soils except the highly porous coarse textured soils. Results from many field experiments found that application of USG @ 80-100 kg ha^{-1} resulted in higher yield advantage of about 400-600 kg rice ha^{-1} over prilled urea. However, in light textured soils, the USG has not found suitable mainly due to leaching of unhydrolyzed urea-N. The use of USGs has not gone into commercial production as yet. In addition to the advantage of N placement, which reduces volatilization losses, placement of such a high amount of urea at a micro-site produces very high concentrations of NH_3. Availability of appropriate machine to place USG in rice fields is the major limitation for their use in Indian agriculture.

7.2.3.3. Sparingly soluble compounds or materials of limited water solubility and plant available forms

Magnesium ammonium phosphate ($MgNH_4PO_4.H_2O$) is the most familiar member of this group of compounds with the general formula $MeNH_4PO_4.XH_2O$, where, Me represents a divalent metals such as Mg, Cu, Co, Mn, Fe and Zn. All of these compounds exhibit only slight solubility in water. The N content of these materials is low (5-9% N), while the P_2O_5 content various from 29 to 45%. Rate of nitrification and release of plant nutrients can be controlled by use of different granule sizes, with the larger one delaying these processes. The solubility of magnesium ammonium phosphate increases as the acidity of soil increases. This material can be used to turf grasses, tree seedlings, ornamentals, citrus, vegetables and field crops.

7.2.3.4. Soluble or relatively water-soluble substances that gradually decompose

The important soluble substances that gradually decompose when applied to soils are (i) Guanylurea sulphate (GUS) containing 37% N, and (ii) Guanylurea phosphate (GUP) containing 27.8%. They are soluble in water but still have slow release properties. They are adsorbed on soil colloids and in that state they mineralize very slowly, particularly under aerobic conditions. Mineralization is apparently much more rapid under anaerobic conditions, such as in flooded rice soils.

7.3. Stabilized N-Fertilizers

Stabilized N-fertilizers are those which are produced by adding some stabilized material to the main N-fertilizer during the production. The N stabilizers are substances such as urease and/or nitrification inhibitors, which when added to the N-fertilizer extend the time of hydrolysis or nitrification processes, respectively so that the N component of the fertilizer remains in the soil in the urea-N or NH_4^+-N forms. The extent of retardation of the hydrolysis or nitrification depends on the amount of stabilized substance added to its main component which may be for a few days to few months. Urease inhibitors are the substances that inhibit hydrolytic action on urea by the enzyme urease, while nitrification inhibitors are the substances that inhibit the biological oxidation of NH_4^+-N to NO_3^--N. In some publications, urease and nitrification inhibitors are designated under slow or controlled release fertilizers.

7.4. Urease Inhibitors

Urease inhibitors are the chemicals that inhibit hydrolytic action of urease enzyme on urea. Upon application to soil, urea is acted on by the enzyme urease, which hydrolyses it to unstable ammonium carbamate (NH_2COONH_4) and then to ammonium carbonate [$(NH_4)_2CO_3$] and finally to ammonium-N (NH_4^+-N) form as shown below:

$$\underset{\text{(Urea)}}{NH_2CONH_2} \xrightarrow[+H_2O]{\text{Urease}} \underset{\text{(Ammonium carbamate)}}{NH_2COONH_4} \xrightarrow[+H_2O]{\text{decomposition}} (NH_4)_2CO_3 \quad \ldots(7.12)$$

$$(NH_4)_2CO_3 \rightarrow 2\ NH_3 + H_2O + CO_2 \quad \ldots(7.13)$$

$$\text{Or, } (NH_4)_2CO_3 \rightarrow 2\ NH_4^+ + CO_3^{2-} \quad \ldots(7.14)$$

The above hydrolysis process is a stepwise conversion mediated by urease enzyme and ammonium carbonate is produced as an intermediate compound during hydrolysis. Urease inhibitors are the substance that helps to reduce the enzymatic breakdown of urea to NH_4^+-N form. They inhibit urea hydrolysis for 2 weeks or more depending on soil and climatic conditions. Since the N use efficiency is very low, especially in puddled soil with the application of urea due to faster rate of hydrolysis. The fertilizer N use efficiency can be improved by reducing the urease activity using a number of compounds that act as urease inhibitors.

7.4.1. Urease enzyme

This enzyme catalyzes the hydrolysis of urea and it occurs universally and is abundant in soils. Large numbers of bacteria, fungi, and actinomycetes in soil possess urease. It is a small group of bacteria known as urea bacteria have the ability to decompose urea. Activity of urease tends to increase with the size of the soil microbial population and increases with organic matter content. The presence of relatively fresh plant residues often results in abundant supplies of urease. Thus, the sources of urease are (i) microorganisms, (ii) seeds of some legume crops, *e.g.* arhar, rajma, soybean, *etc.* and (iii) root exudates. Urease activity is more in the rhizosphere, where microbial activity is high and plant roots excretes more exudates. The activity of urease enzymes are affected by temperature, pH, moisture, ammonia concentration, *etc.*

(i) ***Temperature***: Warm temperature up to 37 °C favour urease activity, but hydrolysis of urea occurs at significant rates at temperature down to 2 °C or below (-20 °C). Urea hydrolysis is generally slow at soil temperature below 10 °C, and as a result NH_4^+ formation is more slow at low temperature.

(ii) ***Moisture***: Soil moisture has less impact on urease enzyme as compared to temperature and pH. The rate of hydrolysis is maximum at soil moisture content in the readily available range to plants.

(iii) ***Ammonia concentration***: Free NH_3 inhibits the enzymatic action of urease. It is evident that some temporary inhibition of urease by free NH_3 will take place after the addition of urea because soil pH in the immediate vicinity of the urea source may reach above 8-9. As a result, high rate of urea create a conditions restrictive to the enzymatic action. Thus, hydrolysis of urea in alkaline condition requires 7 days as compared to 2-3 days in normal soil.

7.4.2. Advantages of Urease Inhibitors

The principal advantages of urease inhibitors are:

(i) A significant reduction in NH_3 volatilization losses, resulting in increase in plant available N;

(ii) An improvement of N-use efficiency from urea and urea-containing fertilizers;

(iii) A reduction of seedling damage; and

(iv) A decrease in the emission of N oxides and nitrous oxide.

7.4.3. Characteristics of Urease Inhibitors

Substances capable of inhibiting urease enzyme should have the following characteristics.

(i) It should be effective at low concentration.

(ii) It should be relatively non-toxic to plants and microorganisms.

(iii) The cost of urease inhibitor must be low or inexpensive.

(iv) It should be compatible with urea.

The prospects for improving the effectiveness of urea through urease inhibitors do not appear as promising as more and more direct alternative means are being developed. Moreover, it increases the cost of production.

7.4.4. Classification of Urease Inhibitors

All the known-urea inhibitors can be grouped into 3 categories:

(i) **First group** of substances inhibits urease activity by blocking essential sulphydril group at active sites on the enzyme. They contain metal ions like mercury (Hg), copper (Cu), silver (Ag), *etc*. Here, Hg is best, but it is not used in soil as it is toxic. The cadmium (Cd) and lead (Pb) are also effective inhibitors. Benzoquinone, quinines, and dihydric phenols when present in quinine form also react with the sulphydryl groups of urease. Heterocyclic S-compounds influence urease similarly by combining with sulphydryl groups.

(ii) **Second group** of inhibitor is structural analogues of urea, such as thiourea, methyl urea, and substituted urea. They have structural similarities to urea and inhibit urease by competing for the same active site on the enzyme. These compounds are ineffective at high urea concentration.

(iii) **Third group** of compounds are those that react with the nickel (Ni)-atom in the urease molecule. Inhibition of urease activity is due theoretically to a tie-up of Ni-atom present in urease enzyme. Hydroxamic acids are non-competitive inhibitors of urease and they are extensively studied known inhibitors. Caprylo-hydroxamix acid is the most potent member of this group.

7.4.5. Common Urease Inhibitors

The important substances or chemical compounds used as urease inhibitors are discussed below:

(i) **N-(butyl) thiophosphoric triamide (NBPT):** Most commonly used urease inhibitor is NBPT [N-(butyl) thiophosphoric triamide) (commercial name **Agrotain®**). It competes for active sites on the urease enzyme and ties up activity for about 10 days, depending on weather conditions. At present, NBPT is the only urease inhibitor of commercial and practical importance in agriculture. It can be injected into molten urea before granulation and applied to the surface of granules or prills in batch or continuous processes.

Agrotain Plus is a dry concentrate containing the urease inhibitor NBPT plus the nitrification inhibitor dicyandiamide (DCD). It reduces volatilization losses of NH_3 as well as N losses from denitrification and leaching.

Super Urea is a granulated urea containing the urease inhibitor NBPT plus the nitrification inhibitor DCD.

(ii) **Phenyl phosphorodiamidate (PPDA):** Phenyl phosphorodiamidate (PPDA) is also acts as urease inhibitor and effectively inhibits urease activity in flooded soils for about 4 days. It can be added with liquid fertilizer containing urea and can be co-granulated with urea.

(iii) **Phenyl Mercury Acetate (PMA):** It is an effective chemical used as urease inhibitor.

(iv) **Other potential urease inhibitors:** Other potential urease inhibitors are **Hydroquinone (HQ)** and **Cyclo-hexyl-phosphoric triamide (CHPT)**.

7.5. Nitrification Inhibitors

Nitrification is the biological oxidation of ammonium to nitrate. It is a two-step process in which the ammonium (NH_4^+) is first converted to nitrite (NO_2^-) and then to nitrate (NO_3^-) as follows:

$$NH_4^+ \xrightarrow[\text{Nitrosomonas}]{[O]} NO_2^- \xrightarrow[\text{Nitrobacter}]{[O]} NO_3^- \quad ...(7.15)$$

If this inorganic N is not taken up by the crop soon after application, then it is likely to be lost through leaching or returned to the atmosphere as a result of denitrification process. In order to reduce the losses of N, the nitrification process must be slowed down. Since nitrification is a biological process, the process can be slowed down by inhibiting the activity of nitrifying (*Nitrosomonas*) bacteria. This can be done by using certain substances that are toxic to the nitrifying bacteria. Nitrification inhibitors are the substances which delay or inhibit the bacterial conversion of NH_4^+ to NO_3^- over a period of time (4-10 weeks). Thus, when they are added to the soils, they temporarily inhibit the process of nitrification, resulting in less amount of NO_3^- formation, which in turn, reduces the loss of N. Furthermore, nitrification inhibitors, by delaying the conversion of NH_4^+ to NO_3^- avoid undesirable high NO_3^- levels in groundwater. Nitrification inhibitors also help to reduce potential losses of N to the atmosphere, allowing more time for the crops to utilize it.

7.5.1. Characteristics of Ideal Nitrification Inhibitor

An ideal nitrification inhibitor should have the following characteristics:

(i) It should ideally be non-toxic to plants, other soil organisms, fish and mammals.

(ii) It should block the conversion of NH_4^+ to NO_3^- by specifically inhibiting *Nitrosomonas* activity.

(iii) It should not interfere with the transformation of nitrite by *Nitrobacter*.

(iv) It should be stable enough for its inhibitory action to last for an adequate period of time usually from several weeks (4-10 weeks).

(v) It should be able to move with the fertilizer or fertilizer solution so that it will be distributed uniformly throughout the soil zone.

(vi) It should be relatively inexpensive so that it can be used on a commercial scale.

7.5.2. Common Nitrification Inhibitors

Many chemicals have been tested in order to search a suitable chemical that have the ability to inhibit nitrification process in soils and thereby more effectively manage the added fertilizer and increase the fertilizer use efficiency. The best known and more effective compounds that are commercially used as nitrification inhibitors are discussed below:

(i) **N-Serve®** [2-Chloro-6(trichloromethyl) pyridine] frequently referred to as **Nitrapyrin**. It is the most important nitrification inhibitor produced commercially. It is prepared by **Dow Chemicals in USA**. Nitrapyrin can delay nitrification by 82% in 14 days. Nitrapyrin can be added to any ammonium fertilizer such as ammonium sulphate, ammonium nitrate, urea, and anhydrous ammonia. However, incorporation of nitrapyrin into conventional N-fertilizers during their production process is difficult due to its high vapour pressure. It is also costly.

(ii) **AM** (2-Amino-4-chloro-6-methyle pyrimidine), which is a substituted pyrimidine. It is prepared by **Mitsui Chemical Corporation in Japan**.

The minimum concentration of N-Serve and AM required to delay nitrification for at least 6 weeks ranges from 0.05-0.2 ppm. Studies showed that when ammonium sulphate was treated with N-Serve and AM @ 2-5 ppm, the nitrification was considerably retarded and even after 30 days, its rate was only 70-76%.

(iii) **Dicyandiamide (NH_2-C=NH-NH-C=N)**

Dicyandiamide (DCD) has been tested as both a nitrification inhibitor and as a slow-release N source. In Japan, it has been added to mixed fertilizers and a product containing urea-form plus 10% DCD (w/w) is manufactured. The DCD is used in both cases to reduce the nitrification process and to increase the content of water soluble N. A fertilizer comprised of urea and DCD in a 4:1 ratio is commercially available in West Germany. Research conducted at Rothamsted Experimental Station, reported that DCD is readily soluble and stable in anhydrous NH_3. Nitrification of anhydrous NH_3 was effectively inhibited for up to 3 months by the addition of 15 kg DCD ha^{-1}. Application of DCD @ 5-10 ppm inhibits the nitrification of NH_4^+-N in fertilizer added to soils. It can delay nitrification by 53% in 14 days.

(iv)**Terrazole (5-Ethoxy-3-trichlonomethyl-1,2,4-thiadiazole):** It is prepared and marketed in USA.

(v) **ATC (4-Amino-1,2,4- triazole hydrochloride):** It is used in Japan, USA and Canada.

(vi) **ST (Sulphathiazole) (2-Sulphanilamido-thiazole)**: It is introduced in Japan.

(vii) **DCS (N-2,5-Dichlorophenyl succinamic acid):** It is identified in Japan and found quite satisfactory inhibitor.

(viii) **DMPP (3,4-dimethylpyrazole phosphate):** It is also used as nitrification inhibitor.

(ix) **Guanyl thiourea (1-Amido-2-thiourea):** This has been patented in Japan and used as a nitrification inhibitor.

(x) **Carbon disulphide (CS_2) and related compounds (Tri-thiocarbonates):** They decompose to release CS_2, have been shown to be good nitrification inhibitors, and their costs may be less than some of the commercially available inhibitors. Compounds like xanthates, which also decompose to release CS_2 is also used as nitrification inhibitor in England and Canada.

(xi) **Other chemicals:** Other chemicals that have been developed and studied as nitrification inhibitors are: (a) Isothiocyanates, (b) Halophenols, (c) Nitroanilines, (d) Haloanilines, (e) Thiourea, (f) ammonium thiosulphate $(NH_4)_2S_2O_3)$, (g) Methionine, (h) Cyanoquanide, (i) Sodium chlorate, (j) Na or K-Azides, (k) Dithane, (l) Dithiocarbamates, *etc.*

(xii) **Naturally occurring substances:** Besides chemical compounds, some naturally occurring substances are also used as nitrification inhibitors. These are easily available low cost materials. A special interest in the use of non-edible oil cakes such as *neem* (*Azadirachta indica*), *mahua* (*Madhuca* *longifolia*), *karanj* (*Pongamia pinnata*) and castor (*Ricinus communis*) arose in the early 1970's when it became known that *neem cake* contain some alkaloids that inhibit the process of nitrification. Commercial nitrification inhibitors like N-Serve and/or AM tried earlier found to be costly for large scale application. Hence, numerous investigations have been made since then to study the potentiality and practicability of non-edible oil cakes as nitrification inhibitors.

Neem coated urea (NCU) has shown to have nitrification inhibiting properties and has been developed in India for commercial production. The major factor responsible for N regulation is nitrification inhibition which is due to presence of some alkaloids in **neem seeds** or **kernel** such as *azadirachtin, nimbin, epinimbin, deacetylnimbin, nimbolinin, nimbidin, nimbidol, salanin, deactylsalanin, etc.* Similarly, **karanj seed** contains *karanjin, glabrin, pogamol, etc.* and **mahua** contains *mowrin.*

Since coating of urea with neem cake was industrially not feasible due to the large volumes involved, a neem oil micro-emulsion (neem seed extract) technique was developed in India. This technique or its modification is currently being used by different fertilizer companies in India. It increased yields by 6-11% in rice and rice-wheat cropping systems.

7.6. New Generation Fertilizers or Smart Fertilizers

In recent years, in order to enhance nutrient use efficiency, **new generation fertilizers** known as **smart fertilizers** are being developed having controlled nutrient release properties. Smart fertilizers may be defined as the fertilizers which works intelligently synchronizing the release of nutrients in tandem with crop demand *(right amount in right time)* having least adverse effects on environment (*minimize pollution*). Like "smart person" having clean, orderly, well dressed, showing a quick-witted intelligence and quick in action, "smart fertilizer" are those fertilizers having low or devoid of impurities (clean), works according to the requirement of the plant and very quick in action. The molecule is water insoluble but has a "smart" feature so that nutrient is released only on demand by the crop. These are designed molecules that allow sustained release of nutrients by a plant-root activated mechanism. The fertilizer molecule functions like a nutrient storehouse providing a continuous nutrient supply throughout the crop growth period.

The development of smart fertilizers based on nanotechnology is a recent phenomenon, with an emphasis on controlled-release and/or carrier/delivery systems to synchronize nutrient availability with the plant demands, thus reducing losses to the environment (Fig. 7.4).

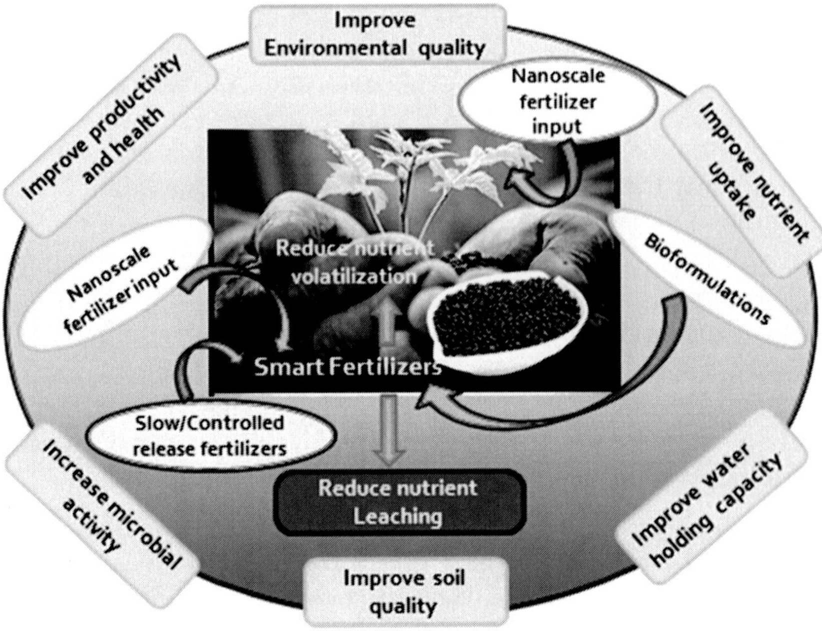

Fig. 7.4. Schematic diagram of smart fertilizer effects in the soil–plant system

7.6.1. Nanofertilizers

Nanofertilizers, as smart fertilizers, are designed to increase nutrient use efficiency and consequently reduce adverse effects on the environment compared to application of conventional mineral fertilizers. Like conventional fertilizers, the nanofertilizers are also fertilizers containing nutrients, in whole or part, of nano-structured formulation(s) that can be delivered to the plants, allowing for efficient uptake or slow release of active ingredients. The development of nanofertilizers could be based on the use of nanomaterials or nanoparticles which can be prepared by coating conventional fertilizers with nanomembranes that facilitate slow and steady release of nutrients. Advancement in nanotechnology has improved ways for large-scale production of nanoparticles which are now used to improve fertilizer formulations for increased uptake in plant cells and by minimizing nutrient loss. Nanofertilizers and nanocomposites are excellent and emerging alternatives to conventional soluble fertilizers. Nutrients are released at a slower rate throughout the crop growth and plants are able to take up most of the nutrients without any waste. Nanofertilizer technology is very innovative that is based on new emerging areas of nanotechnology which has opened up new opportunities to improve nutrient use efficiency, minimize nutrient losses thereby eutrophication and contamination of drinking water and reduce the costs of environmental protection. Nanotechnology has a number of potential benefits ranging from improved food quality and safety to reduced agricultural inputs and improved processing and nutrition.

According to **United States Environment Protection Authority** (USEPA, 2007), nanotechnology may be defined as the understanding and control of matter at dimensions of roughly 1-100 nm, where unique physical properties make novel applications possible. **Nanoparticles** are the materials having one of the dimensions in the size of 1-100 nm. At this nano-scale, the physical, chemical and biological properties of nano-materials differ fundamentally from the properties of individual atoms, molecules, or bulk matter.

In the literature related to nanotechnology application in agriculture, nanofertilizer is used for both materials of a physical diameter between 1 and 100 nm in at least one dimension (*e.g.*, ZnO nanoparticles) and those existing at the bulk scale with more than 100 nm in size but that have been modified with nanoscale materials (*e.g.*, bulk fertilizer coated with nanoparticles).

Nano-science is the science of observing, measuring, and understanding the properties, behaviour, functionality and phenomena of nano-sized objects due to the influence of small dimensions.

The exceptional properties of nanoparticles, such as high ratio of surface area to volume size, high sorption capacity, enhanced physicochemical properties and controlled-release kinetics to targeted sites making them smart delivery system., compared to their bulk counterparts, is now emerging as a promising strategy to promote plant growth and productivity. As a result of their unique properties, nanoparticles may influence metabolic activities of the plant to different degrees compared to conventional materials and have the potential to mobilize native nutrients, such as P, in the rhizosphere.

Nanofertilizers offer an unprecedented opportunity to develop concentrated sources of plant nutrients having higher-absorption rate, utilization efficiency, and minimum losses. Nanofertilizers are being prepared by encapsulating plant nutrients into nanomaterials, employing thin coating of nanomaterials on plant nutrients, and delivering in the form of nano-sized emulsions.

There are three main types of nanofertilizers namely, (i) **nanoscale fertilizer** (synthesized nanoparticles), (ii) **nanoscale additives** (bulk products with nanoscale additives), and (iii) **nanoscale coating or host materials** (product coated with nanopolymer or loaded with nanoparticles).

7.6.2. Mechanisms of Nutrient Uptake from Nanofertilizers

In recent lab scale investigations, it has been reported that nanofertilizers can improve crop productivity by enhancing the rate of seed germination, seedling growth, photosynthetic activity, N metabolism, and carbohydrate and protein synthesis. The major mechanisms of nutrient uptake from nanofertilizers are highlighted below:

(i) The nutrient delivery systems of nanofertilizer are such that they regulate the availability of nutrients in crops through slow or control release mechanisms. Delayed availability of nutrients or consistent supply for extended time periods can be achieved through a number of mechanisms.

(ii) These include semi-permeable coatings for controlled solubility of the fertilizer in water, protein materials, occlusion, chemicals, slow hydrolysis of water-soluble compounds of lower molecular weights, and some other unknown means.

(iii) They could precisely release their active ingredients in synchronization to biological demands of crops. Such a slow delivery of nutrients is associated with the covering or cementing of nutrients with nanomaterials. By taking advantage of slow nutrient delivery, the NUE by crops can be increased because of their consistently long-term delivery of nutrients to plants. For example, nutrients can be released over 40–50 days in a slow release fashion rather than the 4–10 days by the conventional fertilizers.

(iv) Another mechanism is that the nano-pores are taken by plants through the stomatal openings in plant leaves which facilitate nanomaterial uptake and their penetration deep inside leaves. Nanofertilizers can be taken by plants easily through porous leaf or root system thereby higher NUE.

(v) Nanofertilizers have higher transport and delivery of nutrients through plasmodesmata which are nano-sized (50–60 nm) channels between cells. The higher NUE and significantly lesser nutrient losses of nanofertilizers lead to higher productivity (6–17%) and nutritional quality of field crops.

(vi) They are highly reactive with other compounds which facilitate the different metabolic process on plants.

(vii) Another mechanism includes utilization of semi-permeable materials and sensors of chemical or biological origin within the fertilizer (Fig. 7.5). These are advanced materials, whose physical or chemical properties can change in response to an external stimulus such as temperature, pH, and electric or magnetic fields.

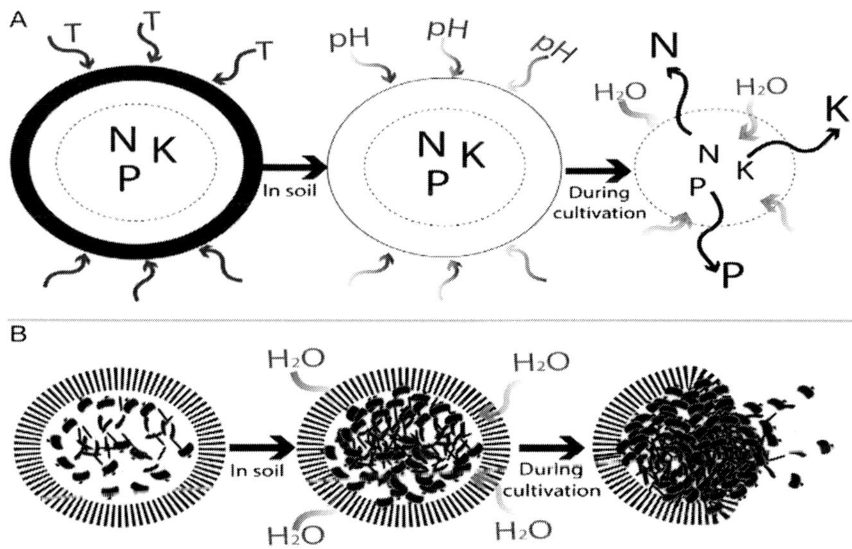

Figure 7.5. Schematic representation of smart delivery systems: (A) advanced polymeric materials, degraded under external stimulus such as temperature, pH, and with water permeability to achieve a slow nutrient release; (B) microorganism encapsulation.

7.6.3. Encapsulation/ Coating/ Supporting Materials for Nanofertilizers

Slow-release nanofertilizers and nanocomposites are suitable alternatives to soluble fertilizers. Nutrients are released at a slower rate during crop growth, thereby reducing the losses of nutrients. Slow release of nutrients in the environments could be achieved by using **zeolites** that are a group of naturally occurring minerals having a honeycomb like layered crystal structure. Zeolite acts as a reservoir for nutrients that are slowly released on demand. Its network of interconnected tunnels and cages can be loaded with N and K, combined with other slowly dissolving ingredients containing P, Ca and micronutrients.

The mineral nutrients required for plant nutrition can be encapsulated inside nanomaterials such as nanotubes or nanoporous materials, coated with a thin protective polymer film, or nanoscale particles or nanodevices or additives (*e.g.*, double hydroxide nanocomposites, urease enzymes, nano-size titanium dioxide, nano-silica particles). Many coating materials can be used to slow nutrient release, including natural materials such as clays and nanoclays (*e.g.*, allophane). Natural clays and nanoclays from montmorillonite and allophanic clays can be used as support materials through encapsulation.

Several studies have shown that urea-based coatings can have variable efficiencies depending on the material used. However, they are highly expensive and in some cases may be toxic and difficult to degrade with potential environmental accumulation. As a consequence, formulations with environmentally safe and biodegradable coating materials are being in progress.

Natural polymeric carbohydrates appear as an alternative to nonbiodegradable materials acting as permeable or impermeable membranes with tiny pores in SRFs. These highly degradable materials have also received attention because of their low cost and low environmental damage due to biodegradability and low accumulation in the environment. However, these polymers may need some modifications to be included into the coating design because of their hydrophilic properties and their weak coating barrier, for example, in the case of starch.

Smart polymeric materials have been applied to smart delivery systems of a wide variety of formulations. A broad range of synthetic materials such as petroleum-based polymers have been used to encapsulate water-soluble fertilizers. Polysulfone, polyacrylonitrile, polyvinyl chloride, polyurethane, and polystyrene are the main materials currently used for coating. While comparing the use of the synthetic polymers polysulfone and polyacrylonitrile and the biodegradable cellulose acetate for the development of slow-release fertilizers, it was observed that physical properties of the coatings can influence the release rate of macronutrients (N, P and K), which are present in the core of the coated fertilizers. It was found that synthetic nondegradable materials had a slower release rate than cellulose acetate-based ones.

Biodegradable Polymers have increasingly been used as substitutes of others polymers in agriculture. They may be divided into two main groups according to their water vapour permeability, namely, degradable synthetic polymers with a small permeability coefficient (***biopols, polylactic acids,*** and ***polycaprolactone***), and ***modified polysaccharides*** with a higher permeability coefficient (alginates, starches, agar). Several studies have reported on the utilization of degradable polymers such as ***chitosan*** and ***chitosan–starch blends*** using sodium tripolyphosphate aqueous solution as the cross-linking agent. They observed that these materials could be used as slow-release N and K fertilizers.

Biodegradable polymers have also been used in ***bioformulations*** which acts as microbial carriers. These carriers protect microbial inoculants from various stresses and prolong shelf life. For example, calcium alginate gel may protect microbial cells with a concomitant increase in shelf life. Sodium alginates are widely used for bioformulations (*i.e.*, bacterial fertilizers) and with pesticides. Despite the low cost and the environmental friendly properties of biodegradable polymers, in many cases the properties of these materials need to be blended with synthetic materials to improve their performance.

Lignocellulosic straw as carrier and coating material for smart fertilizer formulations have been developed. These materials contain lignin, hemicelluloses, and cellulose as well as surface carboxyl, hydroxyl, ether, amino, and phosphate, which enhance its reactivity and physicochemical properties, useful in the preparation of adsorbent materials for the preparation slow-release fertilizers. Moreover, some results have found that wheat straw can be used as reinforcements and/or fillers for non-structural and structural composites.

Nanocomposites are hybrid materials consisting of a continuous (polymer) phase or matrix and a dispersed (nanofiller) phase. The dispersion of a small amount (<10%) of nanomaterial in the polymer matrix can lead to marked improvement in physical and mechanical properties like strengths, pH tolerance, storage stability, heat distortion, break elongation, *etc.* compared with a single polymer matrix.

Currently research is focused to develop nanocomposites to supply essential nutrients through smart delivery system, synchronizing the release of them with the crop uptake, so preventing undesirable nutrient losses to soil (*e.g.*, leaching and volatilization). Coating and cementing of nano and sub-nanocomposites can also be used to regulate the release of nutrients from the fertilizer capsule. Therefore, the future development should focus on materials allowing for nutrient release from nanofertilizers triggered by an environmental condition or simply at specific time.

Smart fertilizers provide significant reduction in rates of fertilizers and yield improvement in an environment-friendly way. Smart fertilizers have been developed for the phosphates (***Smart Phosphates***) and micronutrients (**Smart Micronutrients** fertilizers for zinc, boron, manganese, etc). Smart Phosphate is a replacement for the water-soluble P like DAP, SSP, *etc.*; smart micronutrients can replace the water-soluble micronutrients (zinc sulphate, borax, *etc.*).

With the smart fertilizers, nutrient release is under the control of the plant itself. This makes it especially beneficial in increasing yields since a plant requires different quantities of nutrients at different stages of its life cycle. Elimination of nutrient losses results in reduction of P dosage to between 25% to 50% of current dosages and increase in yields by 10%. For smart micronutrients dosage reduction of 90% is achieved and yields increase by 15-20%. In combination with an environment-friendly nature, the smart phosphate and micronutrients provides a win-win situation for all-farmers, Governments and environment. Smart-release fertilizer technology is globally the first ever and has the potential to bring about far-reaching changes.

7.6.4. Benefits of Nanofertilizers over Conventional Fertilizers

Important benefits of nanofertilizers over conventional fertilizers rely on:

(i) Nanofertilizers exhibit a higher surface area due to their minute particle size and high reactivity of nanofertilzers which increase the bioavailability of nutrients.

(ii) They have slow solubility in water thereby release nutrients slowly for a longer period. Thus, it improves the uptake and nutrient use efficiency in plants.

(iii) In addition, nanofertilizers required in small amount which reduce the cost of transportation and field application.

(iv) An additional major advantage is over accumulation of salt in soil can be minimized as it required in small amount.

(v) Another advantage for using nanofertilizers is that they can be synthesized according to the nutrient requirements of the crops. In this regard, biosensors can be attached to a new innovative fertilizer that controls the delivery of the nutrients according to soil nutrient status, growth period of a crop or environmental conditions.

(vi) Nanofertilizers facilitate the crop plants to fight various biotic and abiotic stresses.

(vii) It is reported that use of nanofertilzers and nanomaterials enhanced the growth and yield in several crops relative to plant treated with conventional fertilizers.

(viii) Nanofertilizers play a significant role on the seed germination and seedlings vigour. Nanofertilizers easily penetrate the epidermis of the seed. As a result, young seedlings get enough available nutrients for their growth and development, thus it helps to increase the yield later on.

(ix) However, the extensive use of nanofertilizers in agriculture may have some limitations which must also be considered. Sometimes it may exhibit inhibitory effects due to excessive amounts of nanofertilizers application. Thus, it is crucial to determine the toxicity/biocompatibility of nanofertilizers.

A comparison of nanotechnology based formulations and conventional fertilizers applications are given in Table 7.1.

7.6.5. IFFCO Nano Nitrogen, Nano Zinc and Nano Copper Fertilizers

The Indian Farmers Fertiliser Cooperative Limited (IFFCO) developed nanotechnology-based products by introducing **Nano Nitrogen**, **Nano Zinc** and **Nano Copper** as alternative to urea, zinc and copper fertilizer, respectively.

Table 7.1. Comparison of nanotechnology based formulations and conventional fertilizers applications

Properties	Nanofertilizers enabled technologies	Conventional technology
Solubility and dispersion of mineral micronutrients	Nano-sized formulation of minerals containing micronutrients may improve solubility and dispersion of insoluble nutrients in soil, reduce soil absorption and fixation, and increase the bioavailability	Less bioavailability to plants due to large particle size and less solubility
Nutrient uptake efficiency	Nano structured formulation might increase fertilizer efficiency and uptake ratio of the soil nutrients in crop production and save fertilizer	Bulk composite is not available for roots resource and decrease efficiency
Controlled release modes	Both release rate and release pattern of nutrients for water may soluble fertilizers might be precisely controlled through encapsulation in envelope forms of semipermeable membranes coated by resin-polymer, waxes, and sulphur	Excess release of fertilizers produce toxicity and destroy ecological balance of soil
Effective duration of nutrient release	Nanostructured formulation can extend effective duration of nutrient supply of fertilizers into soil	Used by the plants at the time of delivery, the rest is converted into insoluble salts in the soil
Loss rate of fertilizer nutrients	Nanostructured formulation can reduce loss rate fertilizer nutrients into soil by leaching and / or leaking	of High loss rate by leaching, rain off, and drift

Source: Cui *et al.* (2010)

These products have been researched and developed indigenously at the IFFCO Nano Biotechnology Research Centre (NBRC), an Advanced Research & Development Centre based at Kalol Unit, Gujarat.

These revolutionary environment-friendly plant nutrition products promise to work 50% better than the conventional fertilizers and could be considered alternative to commercial urea and zinc fertilizers. Presently, these materials are being introduced for on-field trials under controlled conditions with support from ICAR/KVK. If used properly, the nano nitrogen product has the potential to cut the requirement of urea by 50%. While, 10 g ha^{-1} of nano zinc product would be sufficient in place of commercial Zn-fertilizer and would bring the requirement of NPK fertilizer down by 50%. The third product *i.e.* IFFCO nano copper provides both Cu nutrition and protection to the plant. It boosts the plants immunity against harmful pathogens and helps increase the activity of plant growth hormone and improves overall plant growth and development. These nanostructured formulations effectively deliver nutrients to the plants. Other benefits of these nano-products include reduction in the requirement of conventional chemical fertilizer by 50%; increase in crop production by 15-30%; improvement in soil health; reduction in the emission of greenhouse gases.

7.6.6. Environmental and Health Concern of Nanofertilizers

In agriculture, nanotechnology products are being tested for various applications, such as nanoscale sensors for sensing nutrients, nanoscale pesticides, smart and target delivery of nutrients, agronomic fortifications, water purification and nutrient recovery. The benefits of nanofertilizers or nanomaterials are unquestionably opening new approaches towards precision and sustainable agriculture. However, nanoscience and nanotechnology research in agriculture are still at an elementary stage. Following points must be considered carefully.

(i) As it is an infant technology, the ethical and safety issues surrounding the use of nanoparticles in plant productivity are limitless and must be very carefully evaluated before adapting the use of the so-called nanofertilizers in agricultural fields.

(ii) The application of nanofertilizers or nanoparticles as fertilizers should be systematically studied and explored before they could be used in agriculture or farming for a general farm practice.

(iii) Also their limitations should be carefully considered before market implementation.

(iv) The properties of many nanoparticles *viz.*, size, shape, solubility, crystal phase, type of material, and exposure and dosage or concentrations are considered to be of potential risk to environment and human health.

(v) Expert opinions indicate that food products containing nanoparticles available in the market are probably safe to eat, but this is an area that needs to be more actively investigated.

(vi) To address the safety concern detail studies are required to know the impact of nanoparticles within the human body once exposed through nanofood.

(vii) Researchers have to assess and develop proper assessment strategies to assess the impact of nanoparticles and nanofertilizers on biotic and abiotic components of ecosystem.

(viii) Among the various issues, the accumulation of nanomaterials in environment, in edible parts of plants might be the important issues before their use in agriculture.

Chapter 8

Fertiliser Control Order and Fertilisers Specifications

The Fertilizer (Control) Order

The Fertilizer (Control) Order (FCO) is a set of legally enforceable executive orders issued by the Govt. of India under the Essential Commodities Act, 1955 passed by the Parliament. This order came into force in 1957, essentially to regulate the sale, price and quality of fertilizers in India. Amendments are made from time to time as per the exigencies and need of the hour and therefore a large number of amendments were made over a period of time. Consequently, after an overall review of the various provisions of the order, a revised Fertilizer (Control) Order 1985, was issued, which came into effect on September 25, 1985. The FCO consists of five Schedules comprising of different activities as mentioned in **Figure 8.1**, besides forms of registrations and appendices.

Schedule I comprises Part A and B. While Part A contains the specifications of different fertilizers (*e.g.* minimum nutrient content, maximum moisture content, range of physical parameters, maximum levels of impurities/ contaminants, *etc.*), that can be marketed in India (no material not performs the specifications as in the FCO cannot be sold as fertilizer), Part B lists the tolerance limits in plant nutrients and physical parameters for various fertilizers. Schedule II encompasses again two parts, with Part A describing the procedures for drawal of samples of fertilizers, and Part B describing detailed methods of analysis of fertilizers.

Schedule III encompasses four parts namely, Part-A, Part-B, Part-C and Part-D. While Part-A contains specifications of different biofertilizers, Part-B contains tolerance limits of biofertilizers, Part-C describes the procedures for drawal of sample of biofertilizers, Part-D describes the methods of analysis of biofertilizers. Schedule IV also encompasses four parts. Part-A contains specifications of organic fertilizers, Part-B contains tolerance limits of organic fertilizers, Part-C describes the procedures for drawal of sample of organic fertilizers, and Part-D describes the methods of analysis of organic fertilizers. Schedule V also contains of four parts. Part-A contains specifications of non-edible de-oiled cake fertilizers, Part-B contains tolerance limits of non-edible de-oiled cake fertilizers, Part-C describes the procedures for drawal of sample of non-edible de-oiled cake fertilizers, and Part-D describes the methods of analysis of non-edible de-oiled cake fertilizers.

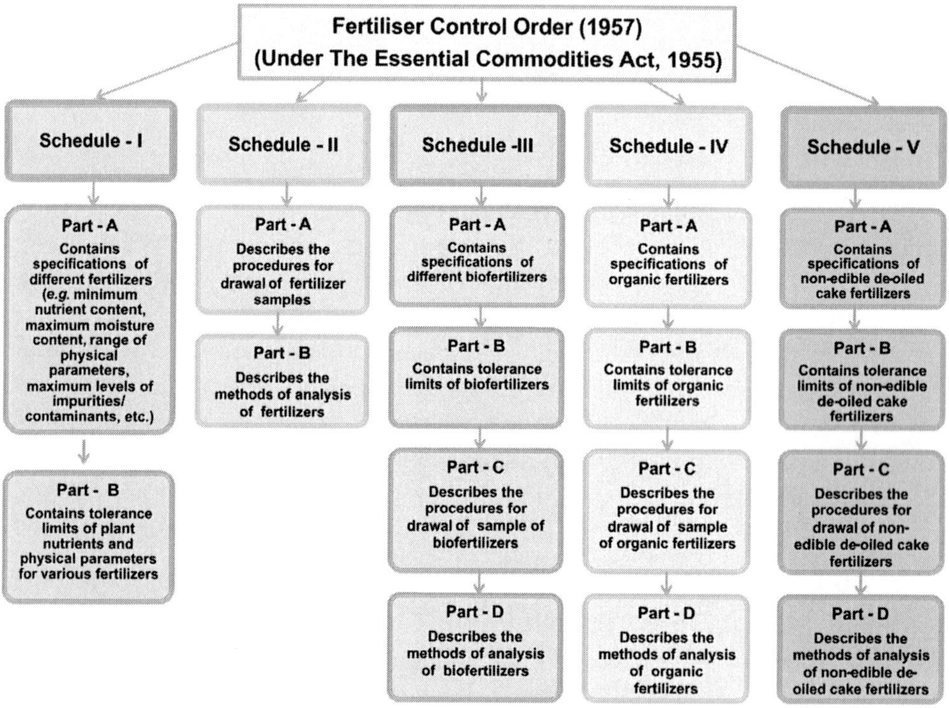

Figure 8.1. Different schedules and activities of FCO
Source: Biswas and Ghosh (2015)

8.1. The Fertiliser Association of India

The Fertiliser Association of India (FAI) is a non-profit, non-trading organization of fertilizer manufacturers, distributors, technologists, plant/ equipment manufacturers, research institutes and others interested in fertilizers. The FAI was set up in 1955 and registered as a company. It is the national representative body of all the fertilizer manufacturers in India, comprising public, private, joint and cooperative sectors. It acts as a bridge between fertilizer industry and Government, and maintains close liaison with international bodies. It publishes the "Fertiliser Statistics', which is regarded as a reference book on Indian fertilizer and agriculture, containing detailed information on production, distribution, consumption, prices and related-statistics from India as well as world.

Specifications of Fertilisers

Specifications of fertilizers as specified by "The Fertiliser (Inorganic, Organic or Mixed) (Control) Order, 1985 (As amended up to February 2019)" are given below:

1.(a) Straight Nitrogenous Fertilisers

1.	**Ammonium Sulphate**	
(i)	Moisture per cent by weight, maximum	1.0
(ii)	Ammoniacal nitrogen, per cent by weight, minimum	20.5
(iii)	Sulphate sulphur (as S), per cent by weight, minimum	23.0
(iv)	Free acidity (as H_2SO_4), per cent by weight, maximum (0.04 for material obtained from by-product ammonia and by-product gypsum)	0.025
(v)	Arsenic as (As_2O_3), per cent by weight, maximum	0.01
2.	**Urea (46% N)**	
(i)	Moisture per cent by weight, maximum	1.0
(ii)	Total nitrogen per cent by weight, (on dry basis), minimum	46.0
(iii)	Biuret per cent by weight, maximum	1.5
(iv)	Particle size – Minimum 80 per cent of the material shall be retained between 1 mm and 2.8 mm IS sieve	
3.	**Ammonium Chloride**	
(i)	Moisture per cent by weight, maximum	2.0
(ii)	Ammoniacal nitrogen per cent by weight, minimum	25.0
(iii)	Chloride other than ammonium chloride (as NaCl) per cent by weight, maximum	2.0
4.	**Calcium Ammonium Nitrate (25% N)**	
(i)	Moisture per cent by weight, maximum	1.0
(ii)	Total ammoniacal and nitrate nitrogen per cent by weight, minimum	25.0
(iii)	Ammoniacal nitrogen per cent by weight, minimum	12.5
(iv)	Calcium nitrate per cent by weight, maximum	0.5

(v)	Particle size – Minimum 90 per cent of the material shall be retained between 1 mm and 4 mm IS sieve	
5.	**Urea Super Granulated**	
(i)	Moisture per cent by weight, maximum	1.0
(ii)	Total nitrogen per cent by weight (on dry basis), minimum	46.0
(iii)	Biuret per cent by weight, maximum	1.5
(iv)	Particle size – Minimum 80 per cent of the material shall be retained between 13.2 mm and 9.5 mm IS sieve	
6.	**Urea (Granular)**	
(i)	Moisture per cent by weight, maximum	1.0
(ii)	Total nitrogen per cent by weight (on dry basis), minimum	46.0
(iii)	Biuret per cent by weight, maximum	1.5
(iv)	Particle size – Minimum 90 per cent of the material shall be retained between 4 mm and 2 mm IS sieve	
7.	**Urea Ammonium Nitrate (32%N) (Liquid)**	
(i)	Total nitrogen (urea, ammoniacal and nitrate) per cent by weight, minimum	32.0
(ii)	Urea nitrogen, per cent weight, maximum	16.5
(iii)	Ammoniacal nitrogen per cent by weight, minimum	7.5
(iv)	Nitrate nitrogen per cent by weight, minimum	7.5
(v)	Specific gravity (at 15 °C)	1.30-1.35
8.	**Neem Coated Urea**	
(i)	Moisture per cent by weight, maximum	1.0
(ii)	Total nitrogen per cent by weight, minimum	46.0
(iii)	Biuret per cent by weight, maximum	1.5
(iv)	Neem oil content soluble in Benzene per cent by weight, minimum	0.035
(v)	Particle size – Minimum 80 per cent of the material shall be retained between 1 mm and 2.8 mm IS sieve	

* ***Note***: The Neem oil used for the manufacturing of Neem Coated Urea may be of the following specifications:

•	Moisture and insoluble matter per cent by weight, maximum	1.0
•	Specific gravity (at 30 °C)	0.85-0.95
•	Iodine value	65-95
•	Saponification value	160-205
•	*Azadirachtin* content, minimum (in ppm)	150

9.	**Urea Briquettes**	
(i)	Moisture per cent by weight, maximum	1.0
(ii)	Total nitrogen per cent by weight (on dry basis), minimum	46.0
(iii)	Biuret per cent by weight, maximum	1.5
(iv)	Particle size – Minimum 90 per cent of the material shall be retained between 5.7 mm and 3.8 mm IS sieve	
10.	**Sulphur Coated Urea**	
(i)	Moisture per cent by weight, maximum	1.0
(ii)	Total nitrogen per cent by weight (on dry basis), minimum	37.0
(iii)	Elemental sulphur (as S) per cent by weight, minimum	17.0
(iv)	Biuret per cent by weight, maximum	1.5
(v)	Particle size – Minimum 90 per cent of the material shall be retained between 8 mm and 2 mm IS sieve	
11.	**Neem Coated Urea (Granular)**	
(i)	Moisture per cent by weight, maximum	1.0
(ii)	Total nitrogen per cent by weight (on dry basis), minimum	46.0
(iii)	Biuret per cent by weight, maximum	1.5
(iv)	Neem oil content soluble in Benzene per cent by weight, minimum	0.035
(v)	Particle size – Minimum 90 per cent of the material shall be retained between 2 mm and 4 mm IS sieve	
12.	**Calcium Cyanamide**	
(i)	Moisture per cent by weight, maximum	1.0
(ii)	Total nitrogen per cent by weight (on dry basis), minimum	19.5
(iii)	Cyanamide nitrogen per cent by weight, minimum	15.0
(iv)	Nitrate nitrogen per cent by weight, maximum	1.5
(v)	Particle size – Not less than 90 per cent of the material shall pass through 4 mm IS sieve and be retained on 1 mm IS sieve	

1 (b). Straight Phosphatic Fertilisers

1. Single Superphosphate (16% P_2O_5 Powdered)

(i)	Moisture per cent by weight, maximum	12.0
(ii)	Available phosphorus (as P_2O_5) per cent by weight, minimum	16.0
(iii)	Water soluble phosphorus (as P_2O_5) per cent by weight, minimum	14.5
(iv)	Sulphate sulphur (as S), per cent by weight, minimum	11.0
(v)	Free phosphoric acid (as P_2O_5) per cent by weight, maximum	4.0

2. Triple Superphosphate

(i)	Moisture per cent by weight, maximum	12.0
(ii)	Total phosphorus (as P_2O_5) per cent by weight, minimum	46.0
(iii)	Water soluble phosphorus (as P_2O_5) per cent by weight, minimum	42.5
(iv)	Free phosphoric acid (as P_2O_5) per cent by weight, maximum	3.0

3. Rock phosphate

(i)	Total phosphorus (as P_2O_5) per cent by weight, minimum	18.0
(ii)	Particle size – Minimum 90 per cent of the material shall pass through 0.15 mm IS sieve and the balance 10 per cent of material shall pass through 0.25 mm IS sieve	

4. Single Superphosphate (16% P_2O_5 Granulated)

(i)	Moisture per cent by weight, maximum	5.0
(ii)	Available phosphorus (as P_2O_5) per cent by weight, minimum	16.0
(iii)	Water soluble phosphorus (as P_2O_5) per cent by weight, minimum	14.5
(iv)	Sulphate sulphur (as S) per cent by weight, minimum	11.0
(v)	Free phosphoric acid (as P_2O_5) per cent by weight, maximum	4.0
(vi)	Particle size – Minimum 90 per cent of the material shall be retained between 1 mm and 4 mm IS sieve	

5. Superphosphoric Acid (70% P_2O_5 Liquid)

(i)	Total phosphorus (as P_2O_5) per cent by weight, minimum	70.0
(ii)	Polyphosphate (as P_2O_5) per cent by weight, minimum	19.0
(iii)	Methanol insoluble matter, per cent weight, maximum	1.0
(iv)	Magnesium as MgO, per cent by weight, maximum	0.5
(v)	Specific gravity (at 24 °C)	1.95-2.00

1(c). Straight Potassic Fertilisers

1. Potassium Chloride (Muriate of Potash)

(i)	Moisture per cent by weight, maximum	0.5
(ii)	Water soluble potassium content (as K_2O) per cent by weight, minimum	60.0
(iii)	Sodium as NaCl per cent by weight (on dry basis), maximum	3.5
(iv)	Particle size – Minimum 65 per cent of the material shall be retained between 0.25 mm and 1.7 mm IS sieve	

2. Potassium Sulphate

(i)	Moisture per cent by weight, maximum	1.5
(ii)	Water soluble potassium (as K_2O) per cent by weight, minimum	50.0

(iii)	Sulphate sulphur (as S) per cent by weight, minimum	17.5
(iv)	Sodium as NaCl per cent by weight, (on dry basis), maximum	2.0
(v)	Total chlorides (as Cl) per cent by weight, maximum	2.5

3. Potassium Schoenite

(i)	Moisture per cent by weight, maximum	1.5
(ii)	Water soluble potassium (as K_2O) per cent by weight (on dry basis), minimum	23.0
(iii)	Magnesium (as MgO) per cent by weight, maximum	11.0
(iv)	Sodium (as NaCl) per cent by weight, maximum	1.5

4. Potassium Chloride (Muriate of Potash) (Granular)

(i)	Moisture per cent by weight, maximum	0.5
(ii)	Water soluble potassium (as K_2O) per cent by weight, minimum	60.0
(iii)	Sodium (as NaCl) per cent by weight, maximum	3.5
(iv)	Magnesium (as $MgCl_2$) per cent by weight, maximum	1.0
(v)	Particle size – Minimum 90 per cent of the material shall be retained between 1 mm and 4 mm IS sieve	

5. Potash Derived from Molasses

(i)	Moisture per cent by weight, maximum	5.0
(ii)	Water soluble potassium (as K_2O), per cent by weight, minimum	14.5

1(d). Straight Sulphur Fertilisers

1. Sulphur 90% (Powder)

(i)	Moisture per cent by weight, maximum	1.0
(ii)	Total elemental sulphur (as S) per cent by weight, minimum	90.0

(*Note*: The product may contain any inert filler material such as Bentonite, *etc.* up to the maximum extent of 10 per cent by weight)

2. Sulphur 90% (Granular)

(i)	Moisture per cent by weight, maximum	0.5
(ii)	Total elemental sulphur (as S) per cent by weight, minimum	90.0
(iii)	Particle size – Minimum 90 per cent of the material shall be retained between 1 mm and 4.0 mm IS sieve.	

(*Note* : The product may contain inert filler material as Bentonite *etc.* up to the extent of 10 per cent by weight)

1(e). N.P. Complex Fertilisers

1. Diammonium Phosphate (18-46-0)

(i)	Moisture per cent by weight, maximum	2.5
(ii)	Total nitrogen (ammoniacal and urea) per cent by weight, minimum	18.0
(iii)	Ammoniacal nitrogen per cent by weight, minimum	15.5
(iv)	Available phosphorus (as P_2O_5) per cent by weight, minimum	46.0
(v)	Water soluble phosphorus (as P_2O_5) per cent by weight, minimum	39.5
(vi)	Particle size – minimum 90 per cent of the material shall be retained between 1 mm and 4 mm IS sieve	

2. Ammonium Phosphate Sulphate (16-20-0)

(i)	Moisture per cent by weight, maximum	1.0
(ii)	Total ammoniacal nitrogen per cent by weight, minimum	16.0
(iii)	Available phosphorus (as P_2O_5) per cent by weight, minimum	20.0
(iv)	Water soluble phosphorus (as P_2O_5) per cent by weight, minimum	17.0
(v)	Sulphate sulphur (as S) per cent by weight, minimum	13.0
(vi)	Particle size – minimum 90 per cent of the material shall be retained between 1 mm and 4 mm IS sieve	

3. Ammonium Phosphate Sulphate (20-20-0)

(i)	Moisture per cent by weight, maximum	1.0
(ii)	Total nitrogen (ammoniacal and urea) per cent by weight, minimum	20.0
(iii)	Ammoniacal nitrogen per cent by weight, minimum	18.0
	Available phosphorus (as P_2O_5) per cent by weight, minimum	20.0
	Water soluble phosphorus (as P_2O_5) per cent by weight, minimum	17.0
(iv)	Sulphate sulphur (as S) per cent by weight, minimum	13.0
(v)	Particle size – minimum 90 per cent of the material shall be retained between 1 mm and 4 mm IS sieve.	

4. Ammonium Phosphate Sulphate Nitrate (20-20-0)

(i)	Moisture per cent by weight, maximum	1.5
(ii)	Total nitrogen (ammoniacal + nitrate) per cent by weight, minimum	20.0
(iii)	Ammoniacal nitrogen per cent by weight, minimum	18.0
(iv)	Available phosphorus (as P_2O_5) per cent by weight, minimum	20.0
(v)	Water soluble phosphorus (as P_2O_5) per cent by weight, minimum	17.0
(vi)	Sulphate sulphur (as S) per cent by weight, minimum	13.0
(vii)	Particle size – minimum 90 per cent of the material shall be retained between 1 mm and 4 mm IS sieve	

5. Nitro Phosphate (20-20-0)

(i)	Moisture per cent by weight, maximum	1.5
(ii)	Total nitrogen (ammoniacal and nitrate) per cent by weight, minimum	20.0
(iii)	Nitrogen in ammoniacal form per cent by weight, minimum	10.0
(iv)	Available phosphorus (as P_2O_5) per cent by weight, minimum	20.0
(v)	Water soluble phosphorus (as P_2O_5) per cent by weight, minimum	12.0
(vi)	Particle size – Minimum 90 per cent of the material shall be retained between 1 mm and 4 mm IS sieve	

6. Urea Ammonium Phosphate (28-28-0)

(i)	Moisture per cent by weight, maximum	1.5
(ii)	Total nitrogen (ammoniacal and urea) per cent by weight, minimum	28.0
(iii)	Ammoniacal nitrogen per cent by weight, minimum	9.0
(iv)	Available phosphorus (as P_2O_5) per cent by weight, minimum	28.0
(v)	Water soluble phosphorus (as P_2O_5) per cent by weight, minimum	24.0
(vi)	Particle size – Minimum 90 per cent of the material shall be retained between 1 mm and 4 mm IS sieve	

7. Urea Ammonium Phosphate (24-24-0)

(i)	Moisture per cent by weight, maximum	1.5
(ii)	Total nitrogen (ammoniacal and urea) per cent by weight, minimum	24.0
(iii)	Ammoniacal nitrogen per cent by weight, minimum	7.5
(iv)	Available phosphorus (as P_2O_5) per cent by weight, minimum	24.0
(v)	Water soluble phosphorus (as P_2O_5) per cent by weight, minimum	20.5
(vi)	Particle size – Minimum 90 per cent of the material shall be retained between 1 mm and 4 mm IS sieve	

8. Urea Ammonium Phosphate (20-20-0)

(i)	Moisture per cent by weight, maximum	1.5
(ii)	Total nitrogen (ammoniacal and urea) per cent by weight, minimum	20.0
(iii)	Ammoniacal nitrogen per cent by weight, minimum	6.5
(iv)	Available phosphorus (as P_2O_5) per cent by weight, minimum	20.0
(v)	Water soluble phosphorus (as P_2O_5) per cent by weight, minimum	17.0
(vi)	Particle size – Minimum 90 per cent of the material shall be retained between 1 mm and 4 mm IS sieve	

9. Mono Ammonium Phosphate (11-52-0)

(i)	Moisture per cent by weight, maximum	1.0
(ii)	Total nitrogen all in ammoniacal form per cent by weight, minimum	11.0
(iii)	Available phosphorus (as P_2O_5) per cent by weight, minimum	52.0
(iv)	Water soluble phosphorus (as P_2O_5) per cent by weight, minimum	44.5
(v)	Particle size – Minimum 90 per cent of the material shall be retained between 1 mm and 4 mm IS sieve	

10. Ammonium Nitrate Phosphate (23-23-0)

(i)	Moisture per cent by weight, maximum	1.5
(ii)	Total nitrogen (ammoniacal and nitrate) per cent by weight, minimum	23.0
(iii)	Nitrogen in ammoniacal form per cent by weight, minimum	13.0
(iv)	Available phosphorus (as P_2O_5) per cent by weight, minimum	23.0
(v)	Water soluble phosphorus (as P_2O_5) per cent by weight, minimum	20.0
(vi)	Particle size – Minimum 90 per cent of the material shall be retained between 1 mm and 4 mm IS sieve.	

11. Ammonium Poly-phosphate (10-34-0) (Liquid)

(i)	Total nitrogen (all as ammoniacal nitrogen) per cent by weight, minimum	10.0
(ii)	Total phosphorus (as P_2O_5) per cent by weight, minimum	34.0
(iii)	Poly-phosphorus (as P_2O_5) per cent by weight, minimum	22.0
(iv)	Specific gravity (at 27 oC)	1.2-1.6
(v)	pH (5% solution)	5.8-6.2

12. Ammonium Phosphate (14-28-0)

(i)	Moisture per cent by weight, maximum	1.5
(ii)	Total nitrogen (ammoniacal and urea) per cent by weight, minimum	14.0
(iii)	Ammoniacal nitrogen per cent by weight, minimum	8.0
(iv)	Available phosphorus (as P_2O_5) per cent by weight, minimum	28.0
(v)	Water soluble phosphorus (as P_2O_5) per cent by weight, minimum	24.0
(vi)	Particle size – Minimum 90 per cent of the material shall be retained between 1 mm and 4 mm IS sieve	

13. N.P.K. 13:33:0:15S

(i)	Moisture per cent by weight, maximum	1.0
(ii)	Ammoniacal nitrogen per cent by weight, minimum	13.0
(iii)	Available phosphorus (as P_2O_5) per cent by weight, minimum	33.0
(iv)	Water soluble phosphorus (as P_2O_5) per cent by weight, minimum	28.5

(v)	Total (elemental + sulphate) sulphur (as S), per cent by weight, minimum	15.0
(vi)	Sulphate sulphur (as S), per cent by weight, maximum	7.5
(vii)	Particle size – Minimum 90 per cent of the material shall be retained between 1 mm and 4 mm IS sieve	

14. Nitrophosphate (24-24-0)

(i)	Moisture per cent by weight, maximum	1.5
	Total nitrogen (ammoniacal and nitrate) per cent by weight, minimum	24.0
(ii)	Nitrogen in ammoniacal form per cent by weight, minimum	13.5
(iii)	Available phosphorus (as P_2O_5) per cent by weight, minimum	24.0
(iv)	Water soluble phosphorus (as P_2O_5) per cent by weight, minimum	20.5
(v)	Particle size – Minimum 90 per cent of the material shall be retained between 1 mm and 4 mm IS sieve	

15. Neem Coated Urea Ammonium Phosphate (28-28-0)

(i)	Moisture percent by weight, maximum	1.5
(ii)	Total nitrogen (ammoniacal and urea) per cent by weight, minimum	28.0
(iii)	Ammoniacal nitrogen per cent by weight, minimum	9.0
(iv)	Available phosphorus (as P_2O_5) per cent by weight, minimum	28.0
(v)	Water soluble phosphorus (as P_2O_5) per cent by weight, minimum	24.0
(vi)	Neem oil content soluble in Benzene per cent by weight, minimum	0.035
(vii)	Particle size – Minimum 90 per cent of the material shall be retained between 1 mm and 4 mm IS sieve	

1(f). N.P.K. Complex Fertilisers

1. Nitrophosphate with Potash (15-15-15)

(i)	Moisture per cent by weight, maximum	1.5
(ii)	Total nitrogen (ammoniacal and nitrate) per cent, minimum	15.0
(iii)	Ammoniacal nitrogen per cent by weight, minimum	7.5
(iv)	Available phosphorus (as P_2O_5) per cent by weight, minimum	15.0
(v)	Water soluble phosphorus (as P_2O_5) per cent by weight, minimum	4.0
(vi)	Water soluble potassium (as K_2O) per cent by weight, minimum	15.0
(vii)	Particle size – Minimum 90 per cent of the material shall be retained between 1 mm and 4 mm IS sieve	

2. N.P.K. (10-26-26)

(i)	Moisture per cent by weight, maximum	1.0
(ii)	Total nitrogen (ammoniacal and urea) per cent by weight, minimum	10.0

(iii)	Ammoniacal nitrogen per cent by weight, minimum	7.0
(iv)	Available phosphorus (as P_2O_5) per cent by weight, minimum	26.0
(v)	Water soluble phosphorus (as P_2O_5) per cent by weight, minimum	22.5
(vi)	Water soluble potassium (as K_2O) per cent by weight, minimum	26.0
(vii)	Particle size – Minimum 90 per cent of the material shall be retained between 1 mm and 4 mm IS sieve	

3. N.P.K. (12-32-16)

(i)	Moisture per cent by weight, maximum	1.5
(ii)	Total nitrogen (ammoniacal and urea) per cent by weight, minimum	12.0
(iii)	Ammoniacal nitrogen per cent by weight, minimum	9.0
(iv)	Available phosphorus (as P_2O_5) per cent by weight, minimum	32.0
(v)	Water soluble phosphorus (as P_2O_5) per cent by weight, minimum	27.5
(vi)	Water soluble potassium (as K_2O) per cent by weight, minimum	16.0
(vii)	Particle size – Minimum 90 per cent of the material shall be retained between 1 mm and 4 mm IS sieve	

4. N.P.K. (22-22-11)

(i)	Moisture per cent by weight, maximum	1.5
(ii)	Total nitrogen (ammoniacal and urea) per cent by weight, minimum	22.0
(iii)	Ammoniacal nitrogen per cent by weight, minimum	7.0
(iv)	Available phosphorus (as P_2O_5) per cent by weight, minimum	22.0
(v)	Water soluble phosphorus (as P_2O_5) per cent by weight, minimum	19.0
(vi)	Water soluble potassium (as K_2O) per cent by weight, minimum	11.0
(vii)	Particle size – Minimum 90 per cent of the material shall be retained between 1 mm and 4 mm IS sieve	

5. N.P.K. (14-35-14)

(i)	Moisture per cent by weight, maximum	1.5
(ii)	Nitrogen in ammoniacal form per cent by weight, minimum	14.0
(iii)	Available phosphorus (as P_2O_5) per cent by weight, minimum	35.0
(iv)	Water soluble phosphorus (as P_2O_5) per cent by weight, minimum	30.0
(v)	Water soluble potassium (as K_2O) per cent by weight, minimum	14.0
(vi)	Particle size – Minimum 90 per cent of the material shall be retained between 1 mm and 4 mm IS sieve	

6. N.P.K. (17-17-17)

(i)	Moisture per cent by weight, maximum	1.5
(ii)	Total nitrogen (ammoniacal and urea) per cent by weight, minimum	17.0
(iii)	Ammoniacal nitrogen per cent by weight, minimum	5.0
(iv)	Available phosphorus (as P_2O_5) per cent by weight, minimum	17.0
(v)	Water soluble phosphorus (as P_2O_5) per cent by weight, minimum	14.5
(vi)	Water soluble potassium (as K_2O) per cent by weight, minimum	17.0
(vii)	Particle size – Minimum 90 per cent of the material shall be retained between 1 mm and 4 mm IS sieve	

7. N.P.K. (14-28-14)

(i)	Moisture per cent by weight, maximum	1.5
(ii)	Total nitrogen (ammoniacal and urea) per cent by weight, minimum	14.0
(iii)	Ammoniacal nitrogen per cent by weight, minimum	8.0
(iv)	Available phosphorus (as P_2O_5) per cent by weight, minimum	28.0
(v)	Water soluble phosphorus (as P_2O_5) per cent by weight, minimum	24.0
(vi)	Water soluble potassium (as K_2O) per cent by weight, minimum	14.0
(vii)	Particle size – Minimum 90 per cent of the material shall be retained between 1 mm and 4 mm IS sieve	

8. N.P.K. (19-19-19)

(i)	Moisture per cent by weight, maximum	1.5
(ii)	Total nitrogen (ammoniacal and urea) per cent by weight, minimum	19.0
(iii)	Ammoniacal nitrogen per cent by weight, minimum	5.5
(iv)	Available phosphorus (as P_2O_5) per cent by weight, minimum	19.0
(v)	Water soluble phosphorus (as P_2O_5) per cent by weight, minimum	16.5
(vi)	Water soluble potassium (as K_2O) per cent by weight, minimum	19.0
(vii)	Particle size – Minimum 90 per cent of the material shall be retained between 1 mm and 4 mm IS sieve	

9. N.P.K. (17-17-17)

(i)	Moisture per cent by weight, maximum	1.5
(ii)	Total nitrogen (ammoniacal and nitrate) per cent by weight, minimum	17.0
(iii)	Ammoniacal nitrogen per cent by weight, minimum	8.5
(iv)	Available phosphorus (as P_2O_5) per cent by weight, minimum	17.0
(v)	Water soluble phosphorus (as P_2O_5) per cent by weight, minimum	17.0
(vi)	Water soluble potassium (as K_2O) per cent by weight, minimum	13.5
(vii)	Particle size – Minimum 90 per cent of the material shall be retained between 1 mm and 4 mm IS sieve	17.0

10. N.P.K. (20-10-10)

(i)	Moisture per cent by weight, maximum	1.5
(ii)	Total nitrogen (ammoniacal and urea) per cent by weight, minimum	20.0
(iii)	Ammoniacal nitrogen per cent by weight, minimum	4.0
(iv)	Available phosphorus (as P_2O_5) per cent by weight, minimum	10.0
(v)	Water soluble phosphorus (as P_2O_5) per cent by weight, minimum	8.5
(vi)	Water soluble potassium (as K_2O) per cent by weight, minimum	10.0
(vii)	Particle size – Minimum 90 per cent of the material shall be retained between 1 mm and 4 mm IS sieve	

11. N.P.K. (15:15:15)

(i)	Moisture per cent by weight, maximum	1.5
(ii)	Total nitrogen (ammoniacal and urea) per cent by weight, minimum	15.0
(iii)	Ammoniacal nitrogen per cent by weight, minimum	12.0
(iv)	Available phosphorus (as P_2O_5) per cent by weight, minimum	15.0
(v)	Water soluble phosphorus (as P_2O_5) per cent by weight, minimum	13.0
(vi)	Water soluble potassium (as K_2O) per cent by weight, minimum	15.0
(vii)	Particle size – Minimum 90 per cent of the material shall be retained between 1 mm and 4 mm IS sieve	

12. N.P.K. (15:15:15:9(S))

(i)	Moisture per cent by weight, maximum	1.5
(ii)	Total nitrogen (ammoniacal and urea) per cent by weight, minimum	15.0
(iii)	Ammoniacal nitrogen per cent by weight, minimum	12.0
(iv)	Available phosphorus (as P_2O_5) per cent by weight, minimum	15.0
(v)	Water soluble phosphorus (as P_2O_5) per cent by weight, minimum	13.0
(vi)	Water soluble potassium (as K_2O) per cent by weight, minimum	15.0
(vii)	Sulphate sulphur (as S) per cent by weight minimum	9.0
(viii)	Particle size – Minimum 90 per cent of the material shall be retained between 1 mm and 4 mm IS sieve	

13. N.P.K. (12:11:18)

(i)	Moisture per cent by weight, maximum	1.5
(ii)	Total nitrogen (ammoniacal and nitrate) per cent by weight, minimum	12.0
(iii)	Ammoniacal nitrogen per cent by weight, minimum	7.0
(iv)	Available phosphorus (as P_2O_5) per cent by weight, minimum	11.0

(v)	Water soluble phosphorus (as P_2O_5) per cent by weight, minimum	8.0
(vi)	Water soluble potassium (as K_2O) per cent by weight, minimum	18.0
(vii)	Magnesium (as Mg) per cent by weight, minimum	1.0
(viii)	Sulphate sulphur (as S) per cent by weight minimum	7.5
(ix)	Total chlorides (as Cl) percent by weight, maximum	1.0
(x)	Particle size – Minimum 90 per cent of the material shall be retained between 1 mm and 4 mm IS sieve	

14. N.P.K. (16:16:16)

(i)	Moisture per cent by weight, maximum	1.5
(ii)	Total nitrogen (ammoniacal and nitrate) per cent by weight, minimum	16.0
(iii)	Ammoniacal nitrogen per cent by weight, minimum	8.0
(iv)	Available phosphorus (as P_2O_5) per cent by weight, minimum	16.0
(v)	Water soluble phosphorus (as P_2O_5) per cent by weight, minimum	12.0
(vi)	Water soluble potassium (as K_2O) per cent by weight, minimum	16.0
(vii)	Particle size – Minimum 90 per cent of the material shall be retained between 1 mm and 4 mm IS sieve	

15. N.P.K. (9:25:25)

(i)	Moisture per cent by weight, maximum	1.5
(ii)	Total nitrogen (ammoniacal and nitrate) per cent by weight, minimum	9.0
(iii)	Ammoniacal nitrogen per cent by weight, minimum	6.0
(iv)	Available phosphorus (as P_2O_5) per cent by weight, minimum	25.0
(v)	Water soluble phosphorus (as P_2O_5) per cent by weight, minimum	21.5
(vi)	Water soluble potassium (as K_2O) per cent by weight, minimum	25.0
(vii)	Particle size – Minimum 90 per cent of the material shall be retained between 1 mm and 4 mm IS sieve	25.0

16. Nitrophosphate with Potash (14-14 21)

(i)	Moisture per cent by weight, maximum	1.5
(ii)	Total nitrogen (ammoniacal and nitrate) per cent by weight, minimum	14.0
(iii)	Ammoniacal nitrogen per cent by weight, minimum	8.0
(iv)	Available phosphorus (as P_2O_5) per cent by weight, minimum	14.0
(v)	Water soluble phosphorus (as P_2O_5) per cent by weight, minimum	9.0
(vi)	Water soluble potassium (as K_2O) per cent by weight, minimum	21.0
(vii)	Particle size – Minimum 90 per cent of the material shall be retained between 1 mm and 4 mm IS sieve	

17. Nitrophosphate with Potash (21-06-13)

(i)	Moisture per cent by weight, maximum	1.5
(ii)	Total nitrogen (ammoniacal and nitrate) per cent by weight, minimum	21.0
(iii)	Ammoniacal nitrogen per cent by weight, minimum	10.5
(iv)	Available phosphorus (as P_2O_5) per cent by weight, minimum	6.0
(v)	Water soluble phosphorus (as P_2O_5) per cent by weight, minimum	4.0
(vi)	Water soluble potassium (as K_2O) per cent by weight, minimum	13.0
(vii)	Particle size – Minimum 90 per cent of the material shall be retained between 1 mm and 4 mm IS sieve	

18. Nitrophosphate with Potash Grade II (15-15-15)

(i)	Moisture per cent by weight, maximum	1.5
(ii)	Total nitrogen (ammoniacal and nitrate) per cent by weight, minimum	15.0
(iii)	Ammoniacal nitrogen per cent by weight, minimum	8.5
(iv)	Available phosphorus (as P_2O_5) per cent by weight, minimum	15.0
(v)	Water soluble phosphorus (as P_2O_5) per cent by weight, minimum	10.0
(vi)	Water soluble potassium (as K_2O) per cent by weight, minimum	15.0
(vii)	Sulphate sulphur (as S) per cent by weight, minimum	3.5
(viii)	Total chloride (as CI) per cent by weight, maximum	3.5
(ix)	Particle size – Minimum 90 per cent of the material shall be retained between 1 mm and 4 mm IS sieve	

19. Nitrophosphate with Potash Grade (15-9-20)

(i)	Moisture per cent by weight, maximum	1.5
(ii)	Total nitrogen (ammoniacal and nitrate) per cent by weight, minimum	15.0
(iii)	Ammoniacal nitrogen per cent by weight, minimum	8.5
(iv)	Available phosphorus (as P_2O_5) per cent by weight, minimum	9.0
(v)	Water soluble phosphorus (as P_2O_5) per cent by weight, minimum	6.0
(vi)	Water soluble potassium (as K_2O) per cent by weight, minimum	20.0
(vii)	Sulphate sulphur (as S) per cent by weight, minimum	3.5
(viii)	Magnesium (as Mg) per cent by weight, minimum	0.5
(ix)	Particle size – Minimum 90 per cent of the material shall be retained between 1 mm and 4 mm IS sieve	

1 (g) Micronutrients

1. Zinc Sulphate Heptahydrate ($ZnSO_4.7H_2O$)

(i)	Matter insoluble in water per cent by weight, maximum	1.0
(ii)	Zinc (as Zn) per cent by weight, minimum	21.0
(iii)	Sulphate sulphur (as S) per cent by weight, minimum	10.0
(iv)	pH (5% solution) not less than	4.0
(v)	Lead (as Pb) per cent by weight, maximum	0.003
(vi)	Cadmium (as Cd) per cent by weight, maximum	0.0025
(vii)	Arsenic (as As) per cent by weight, maximum	0.01

2. Manganese Sulphate

(i)	Manganese (as Mn) content per cent by weight, minimum	30.5
(ii)	Sulphate sulphur (as S) per cent by weight, minimum	17.0
(iii)	Matter insoluble in water per cent by weight, maximum	1.0
(iv)	pH (5% solution) not less than	4.0
(v)	Lead (as Pb) per cent by weight, maximum	0.003
(vi)	Cadmium (as Cd) per cent by weight, maximum	0.0025
(vii)	Arsenic (as As) per cent by weight, maximum	0.01

3. Borax (Sodium Tetraborate) ($Na_2B_4O_7.10H_2O$) for Soil Application

(i)	Content of boron as (B) per cent by weight, minimum	10.5
(ii)	Matter insoluble in water per cent by weight, maximum	1.0
(iii)	pH (3.8% solution)	9.0 - 9.5
(iv)	Lead (as Pb) per cent by weight, maximum	0.003
(v)	Cadmium (as Cd) per cent by weight, maximum	0.0025
(vi)	Arsenic (as As) per cent by weight, maximum	0.01

4. Copper Sulphate ($CuSO_4.5H_2O$)

(i)	Copper (as Cu) per cent by weight, minimum	24.0
(ii)	Sulphate sulphur (as S) per cent by weight, minimum	12.0
(iii)	Matter insoluble in water per cent by weight, maximum	1.0
(iv)	pH (5% solution)	3.0
(v)	Lead (as Pb) per cent by weight, maximum	0.003
(vi)	Cadmium (as Cd) per cent by weight, maximum	0.0025
(vii)	Arsenic (as As) per cent by weight, maximum	0.01

5. Ferrous Sulphate ($FeSO_4.7H_2O$)

(i)	Ferrous iron (as Fe) per cent by weight, minimum	19.0
(ii)	Sulphate sulphur (as S) per cent by weight, minimum	10.5
(iii)	Free acid (as H_2SO_4) per cent by weight, maximum	1.0
(iv)	Matter insoluble in water, per cent by weight, maximum	1.0
(v)	pH (5% solution)	3.5
(vi)	Lead (as Pb) per cent by weight, maximum	0.003
(vii)	Cadmium (as Cd) per cent by weight, maximum	0.0025
(viii)	Arsenic (as As) per cent by weight, maximum	0.01

6. Ammonium Molybdate ($(NH_4)_6Mo7O_{24}.4H_2O$)

(i)	Molybdenum (as Mo) per cent by weight, minimum	52.0
(ii)	Matter insoluble in water per cent by weight, maximum	1.0
(iii)	Lead (as Pb) per cent by weight, maximum	0.003
(iv)	Cadmium (as Cd) per cent by weight, maximum	0.0025
(v)	Arsenic (as As) per cent by weight, maximum	0.01

7. Chelated Zinc as Zn-EDTA

(i)	Appearance – Free flowing crystalline or powder or Tablet	
(ii)	Zinc content (expressed as Zn), per cent by weight minimum in the form of Zn-EDTA	12.0
(iii)	pH (5% solution)	6.0-6.5
(iv)	Lead (as Pb) per cent by weight maximum	0.003
(v)	Cadmium (as Cd) per cent by weight, maximum	0.0025
(vi)	Arsenic (as As) per cent by weight, maximum	0.01

8. Chelated Iron as Fe-EDTA

(i)	Appearance – Free flowing crystalline / powder	
(ii)	Iron content (expressed as Fe), per cent by weight minimum in the form of Fe-EDTA	12.0
(iii)	pH (5% solution)	5.5-6.5
(iv)	Lead (as Pb) per cent by weight, maximum	0.003
(v)	Cadmium (as Cd) per cent by weight, maximum	0.0025
(vii)	Arsenic (as As) per cent by weight, maximum	0.01

9. Zinc Sulphate Mono-hydrate ($ZnSO_4.H_2O$)

(i)	Appearance – Free flowing powder form	
(ii)	Zinc (as Zn) per cent by weight. minimum	33.0
(iii)	Sulphate sulphur (as S) per cent by weight, minimum	15.0
(iv)	Matter-insoluble in water per cent by weight, maximum	1.0
(v)	pH (5% solution) not less than	4.0
(vi)	Iron (as Fe) per cent by weight, maximum	1.0
(vii)	Lead (as Pb) per cent by weight, maximum	0.003
(viii)	Cadmium (as Cd) per cent by weight, maximum	0.0025
(ix)	Arsenic (as As) per cent by weight, maximum	0.01

10. Magnesium Sulphate

(i)	Magnesium (as Mg) per cent by weight, minimum	9.5
(ii)	Sulphate sulphur (as S) per cent by weight, minimum	12.0
(iii)	Matter insoluble in water per cent by weight, maximum	1.0
(iv)	Lead (as Pb) per cent by weight, maximum	0.003
(v)	pH (5% solution)	5.0 - 8.0
(vi)	Lead (as Pb) per cent by weight, maximum	0.003
(vii)	Cadmium (as Cd) per cent by weight, maximum	0.0025
(viii)	Arsenic (as As) per cent by weight, maximum	0.01

11. Boric Acid (H_3BO_3)

(i)	Boron (as B) per cent weight, minimum	17.0
(ii)	Matter insoluble in water per cent by weight, maximum	1.0
(iii)	Lead (as Pb) per cent by weight, maximum	0.003
(iv)	Cadmium (as Cd) per cent by weight, maximum	0.0025
(v)	Arsenic (as As) per cent by weight, maximum	0.01

12. Di-Sodium Octa Borate Tetra Hydrate ($Na_2B_8O_{13}.4H_2O$)

(i)	Boron (as B) per cent weight, minimum	20.0
(ii)	Matter insoluble in water per cent by weight, maximum	1.0
(iii)	Lead (as Pb) per cent by weight, maximum	0.003
(iv)	Cadmium (as Cd) per cent by weight, maximum	0.0025
(v)	Arsenic (as As) per cent by weight, maximum	0.01

13. Di-Sodium Tetra Borate Penta Hydrate (Granular)

(i)	Boron (as B) per cent weight, minimum	14.5
(ii)	Matter insoluble in water per cent by weight, maximum	1.0
(iii)	Lead (as Pb) per cent by weight, maximum	0.003
	Cadmium (as Cd) per cent by weight, maximum	0.0025
(iv)	Arsenic (as As) per cent by weight, maximum	0.01
(v)	Particle size – Minimum 90 per cent of the material shall be retained between 1.4 mm and 5 mm IS sieve	

14. Zinc Sulphate Monohydrate (Granular)

(i)	Zinc (as Zn) per cent by weight, minimum	33.0
(ii)	Sulphate sulphur (as S) per cent by weight, minimum	15.0
(iii)	Iron (as Fe) per cent by weight, maximum	1.0
(iv)	Matter-insoluble in water per cent by weight, maximum	1.0
(v)	pH (5% solution) not less than	4.0
(vi)	Lead (as Pb) per cent by weight, maximum	0.003
(vii)	Cadmium (as Cd) per cent by weight, maximum	0.0025
(viii)	Arsenic (as As) per cent by weight, maximum	0.01
(ix)	Particle size – Minimum 90 per cent of the material shall be retained between 4 mm and 2 mm IS sieve	

15. Zinc Oxide Suspension Concentrate (39.5% Zn)

(i)	Dense suspension concentrate of liquid Zinc	
(ii)	Zinc (as Zn) per cent by weight, minimum	39.5
(iii)	pH (5% solution)	9.0±1
(iv)	Arsenic (as As) per cent by weight, maximum	0.001
(v)	Lead (as Pb) per cent by weight, maximum	0.003
(vi)	Cadmium (as Cd) per cent by weight, maximum	0.0025
(vii)	Specific gravity (at 15 °C)	1.71-1.75

16. Di-Sodium Tetra Borate Penta Hydrate

(i)	Appearance – Free flowing crystalline	
	Boron (as B) per cent weight, minimum	14.5
(ii)	Matter insoluble in water per cent by weight, maximum	1.0
	Arsenic (as As) per cent by weight, maximum	0.001
(iii)	Lead (as Pb) per cent by weight, maximum	0.001
	Cadmium (as Cd) per cent by weight, maximum	0.0025

17. Anhydrous Borax ($Na_2B_4O_7$)

(i)	Appearance – Free flowing crystalline	
(ii)	Boron (as B) per cent weight, minimum	20.5
(iii)	Matter insoluble in water per cent by weight, maximum	0.1
(iv)	Arsenic (as As) per cent by weight, maximum	0.001
(v)	Lead (as Pb) per cent by weight, maximum	0.001
(vi)	Cadmium (as Cd) per cent by weight, maximum	0.0025

18. Concentrated Liquid Calcium 11%

(i)	Calcium (as Ca) per cent by weight, minimum	11.0
(ii)	pH	9.5±1.0
(iii)	Specific gravity (kg L^{-1})	1.3-1.4

19. Zinc Polyphosphate

(i)	Appearance – Free flowing	
(ii)	Zinc (as Zn) per cent by weight, minimum	16.0
(iii)	DTPA soluble zinc (as Zn) per cent by weight, minimum	15.0
(iv)	Water soluble zinc (as Zn) per cent by weight, maximum	1.0
(v)	Magnesium (as Mg) per cent by weight, maximum	3.0
(vi)	Lead (as Pb) per cent by weight, maximum	0.003
(vii)	Arsenic (as As) per cent by weight, maximum	0.01
(viii)	Cadmium (as Cd) per cent by weight, maximum	0.0025

20. Chelated Zinc as Zn-HEDP

(i)	Appearance – Free flowing crystalline	
(ii)	Zinc (as Zn) per cent by weight, minimum in the form of Zn-HEDP	17.0
(iii)	Lead (as Pb) per cent by weight, maximum	0.003
(iv)	Arsenic (as As) per cent by weight, maximum	0.01
(v)	Cadmium (as Cd) per cent by weight, maximum	0.0025

21. Chelated Iron as Fe-HEDP

(i)	Appearance – Free flowing crystalline	
(ii)	Iron (as Fe) per cent by weight, minimum in the form of Fe-HEDP	17.0
(iii)	Lead (as Pb) per cent by weight, maximum	0.003
(iv)	Arsenic (as As) per cent by weight, maximum	0.01
(v)	Cadmium (as Cd) per cent by weight, maximum	0.0025

22. Colemanite

(i)	Boron (as B) per cent by weight, minimum	11.0
(ii)	Sodium (as NaCl) per cent by weight, maximum	0.5
(iii)	Moisture per cent by weight, maximum	6.5
(iv)	Lead (as Pb) per cent by weight, maximum	0.003
(v)	Arsenic (as As) per cent by weight, maximum	0.01
(vi)	Cadmium (as Cd) per cent by weight, maximum	0.0025

23. Boron Ethanolamine

(i)	Boron (as B) per cent by weight, minimum	10.0
(ii)	pH	8.5±1
(iii)	Specific gravity	1.3-1.4
(iv)	Lead (as Pb) per cent by weight, maximum	0.003
(v)	Arsenic (as As) per cent by weight, maximum	0.01
(vi)	Cadmium (as Cd) per cent by weight, maximum	0.0025

1(h) Fortified Fertilisers

1. Boronated Single Superphosphate (16% P_2O_5 Powdered)

(i)	Moisture per cent by weight, maximum	12.0
(ii)	Available phosphorus (as P_2O_5) per cent by weight, minimum	16.0
(iii)	Water soluble phosphorus (as P_2O_5) per cent by weight, minimum	14.5
(iv)	Sulphate sulphur (as S) per cent by weight, minimum	11.0
(v)	Boron (as B) per cent by weight	0.2-0.3
(vi)	Free phosphoric acid (as P_2O_5) per cent by weight, maximum	4.0

2. Zincated Urea

(i)	Moisture per cent by weight, maximum	1.0
(ii)	Total nitrogen (in urea form) per cent by weight, (on dry basis), minimum	43.0
(iii)	Zinc (as Zn) per cent by weight, minimum	2.0
(iv)	Biuret, per cent by weight, maximum	1.5
(v)	Particle Size – Minimum 80 per cent of the material shall be retained between 1 mm and 2.8 mm IS sieve	

3. Zincated Phosphate (Suspension) – for Seed Treatment

(i)	Total phosphorus (as P_2O_5) per cent by weight, minimum	14.0
(ii)	Available phosphorus (as P_2O_5) per cent by weight, minimum	3.0

(iii)	Total zinc (Zn) per cent by weight, minimum	17.5
(iv)	Lead (as Pb) per cent by weight, maximum	0.003
(v)	pH (5% solution)	8±1

4. Boronated NPK Complex (10:26:26)

(i)	Moisture per cent by weight, maximum	1.5
(ii)	Total nitrogen (ammoniacal and urea) per cent by weight, minimum	10.0
(iii)	Ammoniacal nitrogen per cent by weight, minimum	7.0
(iv)	Available phosphorus (as P_2O_5) per cent by weight, minimum	26.0
(v)	Water soluble phosphorus (as P_2O_5) per cent by weight, minimum	22.5
(vi)	Water soluble potassium (as K_2O) per cent by weight, minimum	26.0
(vii)	Boron (as B) per cent by weight	0.2-0.3
(viii)	Particle Size – Minimum 90 per cent of the material shall be retained between 1 mm and 4 mm IS sieve	

5. Boronated NPK Complex (12:32:16)

(i)	Moisture per cent by weight, maximum	1.5
(ii)	Total nitrogen (ammoniacal and urea) per cent by weight, minimum	12.0
(iii)	Ammoniacal nitrogen per cent by weight, minimum	9.0
(iv)	Available phosphorus (as P_2O_5) per cent by weight, minimum	32.0
(v)	Water soluble phosphorus (as P_2O_5) per cent by weight, minimum	27.5
(vi)	Water soluble potassium (as K_2O) per cent by weight, minimum	16.0
(vii)	Boron (as B) per cent by weight	0.2-0.3
(viii)	Particle Size – Minimum 90 per cent of the material shall be retained between 1 mm and 4 mm IS sieve	

6. Boronated Diammonium Phosphate (18:46:0)

(i)	Moisture per cent by weight, maximum	2.5
(ii)	Total nitrogen (ammoniacal and urea) per cent by weight, minimum	18.0
(iii)	Ammoniacal nitrogen per cent by weight, minimum	15.5
(iv)	Available phosphorus (as P_2O_5) per cent by weight, minimum	46.0
(v)	Water soluble phosphorus (as P_2O_5) per cent by weight, minimum	41.0
(vi)	Boron (as B) per cent by weight	0.2-0.3
(vii)	Particle Size – Minimum 90 per cent of the material shall be retained between 1 mm and 4 mm IS sieve	

7. Zincated NPK Complex (10:26:26)

(i)	Moisture per cent by weight, maximum	1.5
(ii)	Total nitrogen (ammoniacal and urea) per cent by weight, minimum	10.0
(iii)	Ammoniacal nitrogen per cent by weight, minimum	7.0
(iv)	Available phosphorus (as P_2O_5) per cent by weight, minimum	26.0
(v)	Water soluble phosphorus (as P_2O_5) per cent by weight, minimum	22.5
(vi)	Water soluble potassium (as K_2O) per cent by weight, minimum	26.0
(vii)	Zinc (as Zn) per cent by weight	0.5
(viii)	Particle Size – Minimum 90 per cent of the material shall be retained between 1 mm and 4 mm IS sieve	

8. Zincated NPK Complex (12:32:16)

(i)	Moisture per cent by weight, maximum	1.5
(ii)	Total nitrogen (ammoniacal and urea) per cent by weight, minimum	12.0
(iii)	Ammoniacal nitrogen per cent by weight, minimum	9.0
(iv)	Available phosphorus (as P_2O_5) per cent by weight, minimum	32.0
(v)	Water soluble phosphorus (as P_2O_5) per cent by weight, minimum	27.5
(vi)	Water soluble potassium (as K_2O) per cent by weight, minimum	16.0
(vii)	Zinc (as Zn) per cent by weight	0.5
(viii)	Particle Size – Minimum 90 per cent of the material shall be retained between 1 mm and 4 mm IS sieve	

9. Boronated Calcium Nitrate

(i)	Total nitrogen (ammoniacal and nitrate) per cent by weight, minimum	14.5
(ii)	Nitrate nitrogen as N per cent by weight, minimum	13.5
(iii)	Water soluble calcium as per cent by weight, minimum	17.0
(iv)	Boron (as B) per cent by weight, minimum	0.2-0.3

10. Boronated Nitrophosphate with Potash (15:15:15)

(i)	Moisture per cent by weight, maximum	1.5
(ii)	Total nitrogen (ammoniacal and nitrate) per cent by weight, minimum	15.0
(iii)	Ammoniacal nitrogen per cent by weight, minimum	7.5
(iv)	Available phosphorus (as P_2O_5) per cent by weight, minimum	15.0
(v)	Water soluble phosphorus (as P_2O_5) per cent by weight, minimum	4.0
(vi)	Water soluble potassium (as K_2O) per cent by weight, minimum	15.0
(vii)	Boron (as B) per cent by weight, minimum	0.2-0.3
(viii)	Particle Size – Minimum 90 per cent of the material shall be retained between 1 mm and 4 mm IS sieve	

11. Zincated DAP (18:46:0)

(i)	Moisture per cent by weight, maximum	2.5
(ii)	Total nitrogen (ammoniacal and urea) per cent by weight, minimum	18.0
(iii)	Ammoniacal nitrogen per cent by weight, minimum	15.5
(iv)	Available phosphorus (as P_2O_5) per cent by weight, minimum	46.0
(v)	Water soluble phosphorus (as P_2O_5) per cent by weight, minimum	39.5
(vi)	Zinc (as Zn) per cent by weight, minimum	0.5
(vii)	Particle Size – Minimum 90 per cent of the material shall be retained between 1 mm and 4 mm IS sieve	

12. Zincated SSP (16% P_2O_5)

(i)	Moisture per cent by weight, maximum	12.0
(ii)	Available phosphorus (as P_2O_5) per cent by weight, minimum	16.0
(iii)	Water soluble phosphorus (as P_2O_5) per cent by weight, minimum	14.5
(iv)	Sulphate sulphur (as S) per cent by weight, maximum	11.0
(v)	Zinc (as Zn) per cent by weight, minimum	0.5
(vi)	Free phosphoric acid (as P_2O_5) per cent by weight, minimum	4.0
(vii)	Particle Size – Minimum 90 per cent of the material shall be retained between 1 mm and 4 mm IS sieve	

13. Boronated NPK Complex (24:24:0)

(i)	Moisture per cent by weight, maximum	1.5
(ii)	Total nitrogen (ammoniacal and nitrate) per cent by weight, minimum	24.0
(iii)	Ammoniacal nitrogen per cent by weight, minimum	13.5
(iv)	Available phosphorus (as P_2O_5) per cent by weight, minimum	24.0
(v)	Water soluble phosphorus (as P_2O_5) per cent by weight, minimum	20.5
(vi)	Boron (as B) per cent by weight, minimum	0.2-0.3
(vii)	Particle Size – Minimum 90 per cent of the material shall be retained between 1 mm and 4 mm IS sieve	

14. Boronated Single Superphosphate (16% P_2O_5) Granular

(i)	Moisture per cent by weight, maximum	5.0
(ii)	Available phosphorus (as P_2O_5) per cent by weight, minimum	16.0
(iii)	Water soluble phosphorus (as P_2O_5) per cent by weight, minimum	14.5
(iv)	Sulphate sulphur (as S) per cent by weight, minimum	11.0
(v)	Boron (as B) per cent by weight, minimum	0.2-0.3
(vi)	Free phosphoric acid (as P_2O_5) per cent by weight, minimum	4.0
(vii)	Particle Size – Minimum 90 per cent of the material shall be retained between 1 mm and 4 mm IS sieve	

15. Zincated Bentonite Sulphur

(i)	Moisture per cent by weight, maximum	1.0
(ii)	Elemental sulphur (as S) per cent by weight, minimum	65.0
(iii)	Zinc (as Zn) per cent by weight, minimum	18.0
(iv)	Particle Size – Minimum 90 per cent of the material shall be retained between 1 mm and 4 mm IS sieve	

16. Zincated Urea Ammonium Phosphate (18:28:0)

(i)	Moisture per cent by weight, maximum	1.5
(ii)	Total nitrogen per cent by weight, minimum	28.0
(iii)	Ammoniacal nitrogen per cent by weight, minimum	9.0
(iv)	Available phosphorus (as P_2O_5) per cent by weight, minimum	28.0
(v)	Water soluble phosphorus (as P_2O_5) per cent by weight, minimum	25.2
(vi)	Zinc (as Zn) per cent by weight, minimum	0.5
(vii)	Particle Size – Minimum 90 per cent of the material shall be retained between 1 mm and 4 mm IS sieve	

17. Zincated NPK (14-35-14)

(i)	Moisture per cent by weight, maximum	1.5
(ii)	Ammoniacal nitrogen per cent by weight, minimum	14.0
(iii)	Available phosphorus (as P_2O_5) per cent by weight, minimum	35.0
(iv)	Water soluble phosphorus (as P_2O_5) per cent by weight, minimum	29.0
(v)	Water soluble potassium (as K_2O) per cent by weight, minimum	14.0
(vi)	Zinc (as Zn) per cent by weight, minimum	0.5
(vii)	Particle Size – Minimum 90 per cent of the material shall be retained between 1 mm and 4 mm IS sieve	

18. Zincated Ammonium Phosphate Sulphate (20-20-0-13)

(i)	Moisture per cent by weight, maximum	1.5
(ii)	Total nitrogen per cent by weight, minimum	20.0
(iii)	Ammoniacal nitrogen per cent by weight, minimum	18.0
(iv)	Available phosphorus (as P_2O_5) per cent by weight, minimum	20.0
(v)	Water soluble phosphorus (as P_2O_5) per cent by weight, minimum	17.0
(vi)	Sulphate sulphur (as S) per cent by weight, minimum	13.0
(vii)	Zinc (as Zn) per cent by weight, minimum	0.5
(viii)	Particle Size – Minimum 90 per cent of the material shall be retained between 1 mm and 4 mm IS sieve	

19. NP 24-24-0 Fortified with Sulphur

(i)	Moisture per cent by weight, maximum	1.5
(ii)	Total nitrogen (ammoniacal and urea) per cent by weight, minimum	24.0
(iii)	Ammoniacal nitrogen per cent by weight, minimum	7.5
(iv)	Nitrogen in the form of urea, maximum	16.5
(v)	Available phosphorus (as P_2O_5) per cent by weight, minimum	24.0
(vi)	Water soluble phosphorus (as P_2O_5) per cent by weight, minimum	20.0
(vii)	Total sulphur (as S) per cent by weight, minimum	8.0
(viii)	Elemental sulphur (as S) per cent by weight, minimum	4.0
(ix)	Sulphate sulphur (as S) per cent by weight, minimum	4.0
(x)	Particle Size – Minimum 90 per cent of the material shall be retained between 1 mm and 4 mm IS sieve	

20. Boronated Ammonium Phosphate Sulphate

(i)	Moisture per cent by weight, maximum	1.5
(ii)	Total nitrogen (ammoniacal and urea) per cent by weight, minimum	20.0
(iii)	Ammoniacal nitrogen per cent by weight, minimum	18.0
(iv)	Available phosphorus (as P_2O_5) per cent by weight, minimum	20.0
(v)	Water soluble phosphorus (as P_2O_5) per cent by weight, minimum	17.0
(vi)	Sulphate sulphur (as S) per cent by weight, minimum	13.0
(vii)	Boron (as B) per cent by weight, minimum	0.2-0.3
(viii)	Particle Size – Minimum 90 per cent of the material shall be retained between 1 mm and 4 mm IS sieve	

21. NPK 14-35-14 Fortified with Zinc and Boron

(i)	Moisture per cent by weight, maximum	1.0
(ii)	Nitrogen in ammoniacal form per cent by weight, minimum	14.0
(iii)	Available phosphorus (as P_2O_5) per cent by weight, minimum	35.0
(iv)	Water soluble phosphorus (as P_2O_5) per cent by weight, minimum	29.0
(v)	Water soluble potassium (as K_2O) per cent by weight, minimum	14.0
(v)	Zinc (as Zn) per cent by weight, minimum	0.5
(vi)	Boron (as B) per cent by weight, minimum	0.3
(vii)	Particle Size – Minimum 90 per cent of the material shall be retained between 1 mm and 4 mm IS sieve. Not more than 5 per cent shall be below 1 mm IS sieve	

22. SSP Fortified with Zinc and Boron (Granular)

(i)	Moisture per cent by weight, maximum	5.0
(ii)	Available phosphorus (as P_2O_5) per cent by weight, minimum	16.0
(iii)	Water soluble phosphorus (as P_2O_5) per cent by weight, minimum	14.5
(iv)	Zinc (as Zn) per cent by weight, minimum	0.5
(v)	Boron (as B) per cent by weight, minimum	0.2
(vi)	Free phosphoric acid (as P_2O_5) per cent by weight, minimum	4.0
(vii)	Particle Size – Minimum 90 per cent of the material shall be retained between 1 mm and 4 mm IS sieve.	

23. Boronated Sulphur (Granular)

(i)	Moisture per cent by weight, maximum	0.5
(ii)	Total elemental sulphur (as S) per cent by weight, minimum	80.0
(iii)	Boron (as B) per cent by weight, minimum	1.2
(iv)	Particle Size – Minimum 90 per cent of the material shall be retained between 1 mm and 4 mm IS sieve.	

Note: The product may contain inert filler material as Bentonite *etc.* up to the extent of 14 per cent by weight, maximum

24. Urea Ammonium Phosphate 20-20-0 fortified with 13% Sulphur

(i)	Moisture per cent by weight, maximum	1.5
(ii)	Total nitrogen (ammoniacal and urea) per cent by weight, minimum	20.0
(iii)	Available phosphorus (as P_2O_5) per cent by weight, minimum	20.0
(iv)	Water soluble phosphorus (as P_2O_5) per cent by weight, minimum	17.0
(v)	Elemental sulphur (as S) per cent by weight, minimum	6.5
(vi)	Sulphate sulphur (as S) per cent by weight, minimum	6.5
(vii)	Particle Size – Minimum 90 per cent of the material shall be retained between 1 mm and 4 mm IS sieve	

25. Zincated SSP (Granular)

(i)	Moisture per cent by weight, maximum	5.0
(ii)	Available phosphorus (as P_2O_5) per cent by weight, minimum	16.0
(iii)	Water soluble phosphorus (as P_2O_5) per cent by weight, minimum	14.5
(iv)	Free phosphoric acid (as P_2O_5) per cent by weight, minimum	4.0
(v)	Sulphate sulphur (as S) per cent by weight, minimum	11.0
(vi)	Zinc (as Zn) per cent by weight, minimum	0.5
(vii)	Particle Size – Minimum 90 per cent of the material shall be retained between 1 mm and 4 mm IS sieve	

26. Elemental Sulphur fortified with Zinc Oxide in microgranule form

(i)	Elemental sulphur (as S) per cent by weight, minimum	67.0
(ii)	Zinc (as Zn) per cent by weight, minimum	14.0
(iii)	Moisture per cent by weight, maximum	2.0
(iv)	Particle Size – Minimum 90 per cent of the material shall be retained between 1 mm and 4 mm IS sieve	

1(i) 100% Water Soluble Complex Fertilisers

1. Potassium Nitrate (13-0-45)

(i)	Moisture per cent by weight maximum	0.5
(ii)	Total nitrogen (all in Nitrate form) per cent by weight, minimum	13.0
(iii)	Water soluble potassium (as K_2O) per cent by weight, minimum	45.0
(iv)	Sodium (as Na) per cent by weight, maximum	1.0
(v)	Total chloride(as Cl) (On dry basis) per cent by weight, maximum	1.5
(vi)	Matter insoluble in water, per cent by weight, maximum	0.5

2. Mono-Potassium Phosphate (0-52-34) (100% Water Soluble)

(i)	Moisture per cent by weight, maximum	0.5
(ii)	Water soluble phosphates (as P_2O_5) per cent by weight, minimum	52.0
(iii)	Water soluble potassium (as K_2O) per cent by weight, minimum	34.0
(iv)	Sodium (as NaCl) per cent by weight (on dry basis), maximum	0.5

3. Calcium Nitrate

(i)	Total nitrogen (ammonium and nitrate) per cent by weight, minimum	15.5
(ii)	Nitrate nitrogen as N per cent by weight, minimum	14.5
(iii)	Water soluble calcium (as Ca) per cent by weight, minimum	18.5
(iv)	Matter insoluble in water per cent by weight, maximum	1.5

4. Potassium Magnesium Sulphate

(i)	Moisture per cent by weight, maximum	0.5
(ii)	Water soluble potassium (as Ca) per cent by weight, minimum	22.0
(iii)	Magnesium as MgO per cent by weight, minimum	18.0
(iv)	Sulphate sulphur (as S) per cent by weight, minimum	20.0
(v)	Total chloride (as Cl) per cent by weight (on dry basis), maximum	2.5
(vi)	Sodium (as NaCl) per cent by weight (on dry basis), maximum	2.0

5. Mono-Ammonium Phosphate (12-61-0) (100% Water Soluble)

(i)	Moisture per cent by weight, maximum	0.5
(ii)	Ammoniacal nitrogen per cent by weight, minimum	12.0
(iii)	Water soluble phosphorus (as P_2O_5) per cent by weight, minimum	61.0
(iv)	Sodium (as NaCl) per cent by weight, maximum	0.5
(v)	Matter insoluble in water per cent by weight, maximum	0.5

6. Urea Phosphate 17:44:0 (100% Water Soluble)

(i)	Moisture per cent by weight, maximum	0.5
(ii)	Total nitrogen (all in urea form) per cent by weight, minimum	17.0
(iii)	Water soluble phosphorus (as P_2O_5) per cent by weight, minimum	44.0
(iv)	Matter insoluble in water per cent by weight, maximum	0.5

References

Arnon, D.I. and Stout, P.R. (1939) An essentiality of certain elements in minute quantity for plants with special reference to copper. *Plant Physiology* **14**, 371–375.

Banerjee, N.K. and Mosier, A.R. (1989) Coated calcium carbide as a nitrification inhibitor in upland and flooded soils. *Journal of the Indian Society of Soil Science* **37**, 306–313.

Basak, B.B. and Biswas, D.R. (2009) Influence of potassium solubilizing microorganism (*Bacillus mucilaginosus*) and waste mica on potassium uptake dynamics by sudan grass (*Sorghum vulgare* Pers.) grown under two Alfisols. *Plant and Soil* **317**, 235-255,

Basak, B.B., Maity, A. and Biswas, D.R. (2020a) Cycling of natural sources of phosphorus and potassium for environmental sustainability. In: *Biogeochemical Cycles: Ecological Drivers and Environmental Impact*, Katerina Dontsova, Zsuzsanna Balogh Brunstad and Gaël Le Roux (Eds.) Geophysical Monograph Series, John Wiley & Sons, Inc., pp. 285-299, doi.org/10.1002/9781119413332.ch14

Basak, B.B., Maity, A., Ray, P., Biswas, D.R. and Roy, S. (2020b) Potassium supply in agriculture through biological potassium fertilizer: a promising and sustainable option for developing countries. *Archives of Agronomy and Soil Science* doi.org/10.1080/03650340.2020.1821191

Basak, B.B., Sarkar, B., Biswas, D.R., Sarkar, S., Sanderson, P. and Naidu, R. (2017) Bio-intervention of naturally occurring silicate minerals for alternative source of potassium: Challenges and opportunities. *Advances in Agronomy* **141**, 115-145.

Birkeland, Kr. (1906) On the oxidation of atmospheric nitrogen in electric arcs. *Transactions of the Faraday Society*. **2** (December): 98, doi:10.1039/tf9060200098.

Biswas, D.R. and Narayanasamy, G. (1998) Direct and residual effectiveness of partially acidulated phosphate rocks as P fertilizer in cowpea-wheat cropping sequence. *Journal of the Indian Society of Soil Science* **46**, 406–412.

Biswas, D.R. and Narayanasamy, G. (2002) Mobilization of phosphorus from rock phosphate through composting using crop residue. *Fertiliser News* **47**, 53–56.

Biswas, D.R. and Narayanasamy, G. (2006) Rock phosphate enriched compost: an approach to improve low-grade Indian rock phosphate. *Bioresource Technology* **97**, 2243–2251.

Brown, T.J., Idoine, N.E., Wrighton, C.E., Raycraft, E.R., Hobbs, S.F., Shaw, R.A., Everett, P., Kresse, C., Deady, E.A. and Bide, T. (2019) World Mineral Production 2014-18. British Geological Survey, Keyworth, Nottingham.

Cui, H.X., Sun, C.J., Liu, Q., Jiang, J. and Gu, W. (2010) Applications of nanotechnology in agrochemical formulation, perspectives, challenges and strategies. In: International Conference on Nanoagri, Sao Pedro, Brazil, pp. 28–33.

Erro, J., Baigorri, R., Yvin, J.C. and Garcia-Mina, J.M. (2011) ^{31}P NMR Characterization and efficiency of new types of water-insoluble phosphate fertilizers to supply plant-available phosphorus in diverse soil types. *Journal of Agricultural and Food Chemistry* **59**, 1900–1908.

Erro, J., Urrutia, O., Baigorri, R., Aparicio-Tejo, P. Irigoyen, I., Torino, F., Mandado, M., Yvin, J.C. and Garcia-Mina, J.M. (2012) Organic complexed superphosphates (CSP): Physicochemical characterization and agronomical properties. *Journal of Agricultural and Food Chemistry* **60**, 2008–2017.

Erro, J., Urrutia, O., Fransisco, S.S. and Garcia-Mina, J.M. (2007) Development and agronomical validation of new fertilizer compositions of high bioavailability and reduced potential nutrient losses. *Journal of Agricultural and Food Chemistry* **55**, 7831–7839.

Eyde, Sam (1909) The Manufacture of Nitrates from the Atmosphere by the Electric Arc—Birkeland-Eyde Process. *Journal of the Royal Society of Arts*. **57**, 568–576.

FAI (2019) Fertiliser Statistics 2018-2019. 64th Edition, The Fertiliser Association of India, FAI House, 10 Shaheed Jit Singh Marg, New Delhi.

FCO (1985) The Fertiliser (Inorganic, Organic or Mixed) (Control) Order, 1985 (As amended up to February 2019). The Fertiliser Association of India, FAI House, 10 Shaheed Jit Singh Marg, New Delhi, 330p.

Gaur, A.C. and Singh, R. (1982) Integrated nutrient supply systems. *Fertiliser News* **27**, 87–98.

Ghosh, A.B. (1984) History and development of agriculture in India with special reference to soil science during 1800-1950, In *Soil Science in India*. Indian Society of Soil Science, New Delhi, pp.14-28.

Hammond, L.L., Chien, S.H. and Mokwunye, A.U. (1986) Agronomic value of unacidulated and partially acidulated phosphate rocks indigenous to the tropics. *Advances in Agronomy* **40**, 89–140.

IFDC (1986) *Technical Bulletin (T-31)*. International Fertiliser Development Centre, Muscle Shoals, Alabama, USA.

Indian Minerals Yearbook (2019a) *Apatite and Rock Phosph*ate. Part-III: Mineral Reviews, (Advance Release), 58th Edition, Government of India, Ministry of Mines, Indian Bureau of Mines, Indira Bhavan, Civil Lines, Nagpur, 1, 1–15.

Indian Minerals Yearbook (2019b) *Potash*. In *Part- III: Mineral Reviews*, (Advance Release), 58th Edition, Government of India, Ministry of Mines, Indian Bureau of Mines, Indira Bhavan, Civil Lines, Nagpur, 23, 1–5.

Indian Minerals Yearbook (2019c) *Minor Minerals, 30.2 Bentonite*. Part-III: Mineral Reviews, (Advance Release), 58th Edition, Government of India, Ministry of Mines, Indian Bureau of Mines, Indira Bhavan, Civil Lines, Nagpur, 30, 2.1–2.4.

Kundu, S., Adhikari, T., Biswas, A.K., Tarafdar, J.C., Goswami, A. and Subba Rao, A. (2010) *Technical Bulletin*, Indian Institute of Soil Science, Bhopal, pp. 1–46.

Liebig , J. (1840) "*Die organische Chemie in ihrer Anwendung auf Agricultur und Physiologie*", *raunschweig* pp. 165 and 180.

Mamata Begum, Narayanasamy, G. and Biswas, D.R. (2004) Phosphorus supplying capacity of phosphate rocks as influenced by compaction with water-soluble P fertilizers. *Nutrient Cycling in Agroecosystems* **68**, 73–84.

McClellan, G.H. and Lehr, J.R. (1969) Crystal Chemical Investigation of Natural Apatites. *American Mineralogy* **54**, 1374-1391.

Memon, Y.M., Fergus, I.F., Hughes, J.D. and Page, D.W. 1988. Utilization of nonexchangable soil potassium in relation to soil types, plant species and stage of growth. *Australian Journal of Soil Research* **26**, 489-496.

Mengel, K. and Kirkby, E.A. (2001) *Principles of Plant Nutrition*, 5th edition, Kluwer, Dordrecht.

Muralidharudu, Y., Sammi Reddy, K., Mandal, B.N., Subba Rao, A., Singh, K.N. and Sonekar, S. (2011) *GIS Based Soil Fertility Maps of Different States of India*. All India Coordinated Research Project on Soil Test Crop Response, Indian Institute of Soil Science, Bhopal, 224p.

Narayanasamy, G. and Biswas, D.R. (1998) Phosphate rocks of India–Potentialities and Constraints. *Fertiliser News* **43**, 21–28 and 31–32.

Narayanasamy, G., Arora, B.R., Biswas, D.R. and Khanna, S.S (2009) Fertilizers, Manures and Biofertilizers. In: *Fundamentals of Soil Science.* N.N. Goswami, R.K. Rattan, G. Dev, G. Narayanasamy, D.K. Das, S.K. Sanyal, D.K. Pal and D.L.N. Rao (Eds.), Indian Society of Soil Science, New Delhi, pp. 579–622.

Pasricha, N.S. and Sarkar, A.K. (2009) Secondary nutrients. In: *Fundamentals of Soil Science.* N.N. Goswami, R.K. Rattan, G. Dev, G. Narayanasamy, D.K. Das, S.K. Sanyal, D.K. Pal and D.L.N. Rao (Eds.), Indian Society of Soil Science, New Delhi, pp. 449–460.

Prud'homme, M. (2018) Medium-term outlook for fertilizers and raw materials global supply–2018-2022. Paris, France, International Fertilizer Association, 114p.

Reddy, K.S., Narayanasamy, G., Rao A.S. (2000) Thermally promoted phosphate rocks-iii. direct and residual effects in a Typic Ustochrept. *Journal of the Indian Society of Soil Science* **48**, 320–325.

Roy, T., Biswas, D.R., Datta, S.C., Sarkar, S. and Biswas, S.S. (2018) Citric acid loaded nano clay polymer composite for solubilization of Indian rock phosphates: A step towards sustainable and phosphorus secure future. *Archives of Agronomy and Soil Science* **64**, 1564–1581.

Sharpley, A.N. (1989) Relationship between soil potassium forms and mineralogy. *Soil Science Society of America Journal* **64**, 87–98.

Sparks, D.L. and Huang, P.M. (1985) Release of soil potassium by weathering reactions. In: Munson RD (ed) *Potassium in agriculture.* American Society of Agronomy, Madison, WI, pp. 202–276.

Sparks, D.L. (1987) Potassium dynamics in soils. *Advance in Soil Science* **6**, 1–63.

Sparks, D.L. (2000) Bioavailability of soil potassium, D-38-D-52. In: Sumner ME (ed) *Handbook of soil science.* CRC, Boca Raton, FL.

Sureshkumar, P. and Sandeep, S (2015) Secondary Nutrients in Soils and their Management. In *Soil Science: An Introduction.* R.K. Rattan, J.C. Katyal, B.S. Dwivedi, A.K. Sarkar, Tapas Bhattacharyya, J.C. Tarafdar and S.S. Kukal (Eds.), Indian Society of Soil Science, New Delhi, pp. 601–621.

Tandon, H.L.S. (2012) Fertilizer management: Balance-efficiency-profitability. Fertilizer development and consultation organization, New Delhi, 187p.

Tisdale, S.L., Nelson, W.L., Beaton, J.D. and Havlin, J.L. (1997) *Soil Fertility and Fertilizers* (Fifth edition), Second Indian Reprint, Prentice Hall of India Ltd, New Delhi.

U.S. Geological Survey (2020) Mineral commodity summaries 2020: United States Geological Survey, 200p, https://doi.org/10.3133/ mcs2020.

US Environmental Protection Agency (2007) Nanotechnology White Paper. Report, United States Environmental Protection Agency 100/B-07/001. Washington, DC, USA.

USGS (2020) Minerals Yearbook - Phosphate Rock (Advance Release). United States Department of the Interior and United States Geological Survey.

Zhan, F., Liu, M., Guo, M. and Wu, L. (2004) Preparation of superabsorbent polymer with slow release phosphate fertilizer. *Journal of Applied Polymer Science* **92**, 3417–3421.

Further Suggested Reading

Barber, S.A. (1995) *Soil Nutrient Bioavailability: A Mechanistic Approach.* Second Edition. Wiley Interscience, New York, 398p.

Collings, G.H. (1955) *Commercial Fertilizers: Their Sources and Use.* Fifth Edition, Tata McGraw-Hill Publishing Company Ltd, Bombay and New Delhi, 522p.

Fertilizer Manual (1979) International Fertilizer Development Center (IFDC) and United Nations Industrial Development Organization (UNIDO), 218p.

Finck, A. (1992) Fertilizers and their efficient use. In: *World Fertilizer Use Manual.* International Fertilizer Industry Association, Paris, France, 125p.

Goswami, N.N., Rattan, R.K., Dev, G., Narayanasamy, G., Das, D.K., Sanyal, S.K., Pal, D.K. and Rao, D.L.N. (Editors) (2009) *Fundamentals of Soil Science.* Second Edition. Indian Society of Soil Science, New Delhi, 728p.

Havlin, J.L, Tisdale, S.L., Nelson, W.L. and Beaton, J.D. (2013) *Soil Fertility and Fertilizers: An Introduction to Nutrient Management.* Eighth Edition. Prentice Hall, 528p.

Marschner, H. (1995) *Mineral Nutrition of Higher Plants.* Academic Press, Orlando, Florida, USA, 674p.

Mengel, K. and Kirkby, E.A. (1982) *Principles of Plant Nutrition.* International Potash Institute, Bern, Switzerland, 689p.

Olson, R.A., Army, T.J., Hanway, J.J. and Kilmer, V.J. (Editors) (1971) *Fertilizer Technology and Use.* Soil Science Society of America, Inc., Wisconsin, Madison, USA, 611p.

Rattan, R.K., Katyal, J.C., Dwivedi, B.S., Sarkar, A.K., Bhattacharyya, Tapas, Tarafdar, J.C. and Kukal, S.S. (Editors) (2015) *Soil Science: An Introduction.* Indian Society of Soil Science, New Delhi.

Sauchelli, Vincent (Editor) (1964) *Fertilizer Nitrogen: Its Chemistry and Technology*. Reinhold Book Corporation, New York.

Yawalkar, K.S., Agarwal, J.P. and Bokde, S. (1996) *Manures and Fertilisers* (Eighth Revised Edition). Agri-Horticultural Publishing House, Nagpur.

Subject Index

M

N

O

P

R

S

T

U

V

W

Z